Challenge Problems

Algebra 2

Explorations and Applications

Challenge Problems includes additional challenging exercises
and problems for each lesson, plus chapter review exercises and
extension problems. Answers to all exercises and problems are
provided at the end of the book.

McDougal Littell

A Houghton Mifflin Company

Evanston, Illinois

Boston   Dallas

About Challenge Problems

Challenge Problems was designed to be used in conjunction with the student textbook. It contains a set of additional challenging exercises for every section in the textbook, as well as a set to be used at the conclusion of each chapter. These exercises require students to use higher-order thinking skills, especially those exercises marked with a star and the extension exercises in the chapter challenge sets. Answer to all exercises are provided in the end of the book.

Acknowledgment

Challenge Problems was written by John A. Graham, Mathematics Teacher and Computer Coordinator of the Upper School, Buckingham Browne and Nichols School, Cambridge, Massachusetts

ISBN: 0-395-83590-9

123456789 - HS - 00 99 98 97 96

CONTENTS

	Page
Chapter 1 Challenge Sets	**1**
Chapter 2 Challenge Sets	**7**
Chapter 3 Challenge Sets	**14**
Chapter 4 Challenge Sets	**20**
Chapter 5 Challenge Sets	**26**
Chapter 6 Challenge Sets	**33**
Chapter 7 Challenge Sets	**40**
Chapter 8 Challenge Sets	**47**
Chapter 9 Challenge Sets	**54**
Chapter 10 Challenge Sets	**64**
Chapter 11 Challenge Sets	**70**
Chapter 12 Challenge Sets	**77**
Chapter 13 Challenge Sets	**83**
Chapter 14 Challenge Sets	**89**
Chapter 15 Challenge Sets	**96**
Answers	**A1**

Challenge Set 1 ···

FOR USE WITH SECTION 1.1

The table below gives the total annual expenditures (in billions of dollars) for elementary and secondary education in the United States in selected years. Use the data in Exercises 1–6.

School Year	1949–1950	1959–1960	1969–1970	1979–1980	1989–1990
Expenditure	5.8	15.6	40.7	96.0	212.1

1. List the increases in the data from year to year and the growth factors from year to year.

2. Describe a method, based on values like the two sets of values that you found in Exercise 1, that will decide whether a constant growth model or a proportional growth model will make more accurate predictions.

3. Use the two sets of data you found in Exercise 1 to estimate the total annual expenditure for elementary and secondary education in 1939–40. Does one of the two figures seem more likely than the other? If so, why?

4. Describe a method for estimating the expenditure on elementary and secondary education in years like 1964–1965 or 1974–75, based on the constant growth model, and use this method to estimate the expenditure in 1964–1965.

5. Describe a method for estimating the expenditure on elementary and secondary education in years like 1964–1965 or 1974–75, based on the proportional growth model, and use this method to estimate the expenditure in 1974–1975.

6. Another way to find an average growth figure is to subtract the first figure from the last and divide by 4. Find the average growth by this method.

7. Suppose a, b, c, d, and e are 5 data items like the expenditure figures above, and suppose these numbers form an increasing sequence. Show algebraically that the method of finding an average growth figure suggested in Exercise 6 gives the same answer you would get by averaging the annual growth figures.

In Exercises 8 and 9, use the table below, which gives the winning distance (to the nearest foot) of the women's javelin throw in the Olympics held between 1972 and 1992.

Year	1972	1976	1980	1984	1988	1992
Distance (ft)	210	216	224	228	245	224

8. Find the average increase in the winning distance. Find the average growth factor of the distances. Describe how you dealt with the last two figures, which show a decrease in distance, in each case.

★ **9.** *Writing* Describe two methods for finding the average change (up or down) in a set of data such as the distances in the table.

Challenge Problems, ALGEBRA 2: EXPLORATIONS AND APPLICATIONS

Challenge Set 2 ···

1. The table below shows sales at a small retail store for 3 successive years, in thousands of dollars. Find a constant growth model and an exponential growth model for the data. If you were the store manager, which model would you prefer?

Year	1993	1994	1995
Sales ($1000)	248	332	248

The table below gives the annual *percent change* in the Consumer Price Index (CPI) in the years 1986–1993. (Each figure represents the change from the preceding year.) Use this table in Exercises 2–5.

Year	1986	1987	1988	1989	1990	1991	1992	1993
CPI (% change)	1.9	3.6	4.1	4.8	5.4	4.2	?	?

2. The average percent change in the CPI was 3.75 between 1985 and 1993, and the two percent changes in the CPI in 1992 and 1993 were equal. What was this common percent change?

3. The CPI stood at 130.7 in 1990. Use this figure to construct an exponential model for the CPI, using as the variable years after 1990. Use your model to predict the value of the CPI in the year 1985. What value did you use for x?

4. Construct a linear model for the data by computing the actual values of the CPI in each of the years 1990 through 1993. Use your model to predict the value of the CPI in the year 1985.

5. *Writing* The actual value of the CPI in 1985 was 107.6. Which of the models you found in Exercises 3 and 4 gives the better estimate of this value? What conclusions can you draw from your answer?

6. The fundraising office of a charity raised $155,000 in 1990. In 1994 the office raised $271,000.

   a. Suppose the growth was the same for each of the years 1991 through 1994. What was this growth? Use your answer to construct a linear model based on the data. Use your model to predict the amount that will be raised in 1998.

   b. Suppose the growth *factor* was the same for each of the years 1991 through 1994. Find this growth factor. Use the growth factor to construct an exponential model based on the data. (*Hint*: If the growth factor is represented by x, then $155,000x^4 = 271,000$. Note that $x^4 = (x^2)^2$.) Use your model to predict the amount that will be raised in 1998.

★ **7.** Suppose the average growth factor for the annual profit of a manufacturer *for every 4-year period* is represented by k, and suppose the manufacturer made a profit of $4.2 million in 1980. Write an equation, using x to represent number of years after 1980, for the profit y made by the manufacturer. (*Hint*: Use a method suggested by Exercise 6(b).)

Challenge Problems, ALGEBRA 2: EXPLORATIONS AND APPLICATIONS

Challenge Set 3 ···

FOR USE WITH SECTION 1.3

Solve each matrix equation.

1. $\begin{bmatrix} x & -3 \\ z+1 & 0 \end{bmatrix} = \begin{bmatrix} 5-x & 3y \\ 4 & 2w+7 \end{bmatrix}$

2. $\begin{bmatrix} a & 5 \\ -2 & d \end{bmatrix} + \begin{bmatrix} a-3 & b+2 \\ 5c & -1 \end{bmatrix} = \begin{bmatrix} 7 & 1 \\ -12 & 6 \end{bmatrix}$

In Exercises 3–8, use the matrices A, B, and C below. Find the unknown matrix X satisfying each equation.

$$A = \begin{bmatrix} 5 & -1 \\ 2 & 0 \\ -3 & 2 \end{bmatrix} \qquad B = \begin{bmatrix} 3 & 1 \\ 0 & -6 \\ 5 & 2 \end{bmatrix} \qquad C = \begin{bmatrix} 7 & -5 \\ -2 & 6 \\ 1 & 4 \end{bmatrix}$$

3. $A + X = B$

4. $2C - X = A$

5. $3X = A + C$

6. $2X - B = C$

7. $A - 4X = B$

8. $5X + 2B = 3C$

9. Every 2×2 matrix $\begin{bmatrix} a & b \\ c & d \end{bmatrix}$ in which $ad - bc \neq 0$ defines a parallelogram in the plane by means of the following rule:

$\begin{bmatrix} a & b \\ c & d \end{bmatrix}$ corresponds to

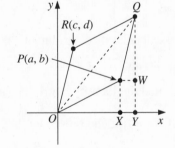

a. In the diagram, the dotted segments $\overline{PX}$ and $\overline{QY}$ are vertical, and the dotted segment $\overline{PW}$ is horizontal. Explain how you know that $PW = c$ and $QW = d$.

b. Use the result of part (a) to find the coordinates of Q in the diagram, in terms of a, b, c, and d.

c. Show that the area of parallelogram $OPQR$ is $ad - bc$, using the following method:

(1) Find the area of triangle OYQ and subtract the areas of triangles OXP and PQW, as well as the area of rectangle $XYWP$. This gives you the area of triangle OPQ.

(2) Multiply your result by 2 to get the area of the parallelogram.

★ **10.** For each pair of dimensions m and n, the *zero matrix O* is the $m \times n$ matrix with every entry equal to 0. Given a 2×2 matrix X, is there a 2×2 matrix Y such that $X + Y = O$. If so, describe its entries in terms of the entries of X. If not, explain why not.

Challenge Set 4 ···

Use the matrices *A, B, C*, and *D* in Exercises 1–4.

$$A = \begin{bmatrix} -5 & 2 \\ 3 & -1 \end{bmatrix} \qquad B = \begin{bmatrix} 2 & 0 \\ 0 & 3 \end{bmatrix} \qquad C = \begin{bmatrix} 4 & 0 \\ 0 & 4 \end{bmatrix} \qquad D = \begin{bmatrix} 3 & -2 \\ 0 & 6 \end{bmatrix}$$

1. Find the products *AB, BA, AC, CA, AD*, and *DA*. Are any of these equal? If so, which ones?

2. Find the products *BC, CB, BD*, and *DB*. Are any of these equal? If so, which?

3. Find the products *CD* and *DC*. Are these equal?

★ 4. On the basis of your results in Exercises 1–3, make a conjecture about a condition on the entries of a 2×2 matrix *X* that guarantees that *X* commutes with every other 2×2 matrix. That is, $XW = WX$ for any 2×2 matrix *W*.

   Prove your conjecture using $W = \begin{bmatrix} a & b \\ c & d \end{bmatrix}$.

5. The number $ad - bc$ has a special association with the matrix $X = \begin{bmatrix} a & b \\ c & d \end{bmatrix}$.

   The number $ad - bc$ is called the *determinant* of the matrix *X* and is often

   denoted det(*X*). Suppose $Y = \begin{bmatrix} p & q \\ r & s \end{bmatrix}$.

   a. Find the matrix *XY* and calculate its determinant.

   b. Find det(*Y*). Check whether $\det(XY) = \det(X) \cdot \det(Y)$ for all 2×2 matrices *X* and *Y*.

6. The matrix $I = \begin{bmatrix} 1 & 0 \\ 0 & 1 \end{bmatrix}$ is called the identity matrix because $IX = X = XI$

   for any 2×2 matrix *X*. If $XY = I$, it can be shown that $YX = I$, and *Y* is

   called the *multiplicative inverse* of *X*. If $X = \begin{bmatrix} a & b \\ c & d \end{bmatrix}$ then

   $Y = \dfrac{1}{\det(X)} \begin{bmatrix} d & -b \\ -c & a \end{bmatrix}$, provided $\det(X) \neq 0$. Find the inverse *Y* of each matrix

   X, and verify that $XY = I$.

   a. $X = \begin{bmatrix} -2 & 3 \\ -5 & 7 \end{bmatrix}$ **b.** $\begin{bmatrix} -4 & 5 \\ 2 & -3 \end{bmatrix}$ **c.** $\begin{bmatrix} 2 & 4 \\ 4 & 6 \end{bmatrix}$

7. If $\det(X) = 0$ for some 2×2 matrix *X*, then *X* has no multiplicative inverse. Find two such matrices *X*.

8. The determinant of a 3×3 matrix $\begin{bmatrix} a & b & c \\ d & e & f \\ g & h & k \end{bmatrix}$ is the number $aek + bfg +$

   $cdh - gec - hfa - kdb$, which is useful in solving linear systems. Describe a pattern, based on the "geometry" of the matrix, that might help you remember this formula.

Challenge Set 5 ···

1. Suppose a basketball player has a free-throw percentage of 60%, and suppose that this player has a chance to shoot a free throw, getting a chance to shoot a second free throw if she makes the first one.

   a. Describe a simulation of the situation.

   b. Repeat your simulation 25 times. Calculate the probability of making no free throw, exactly one free throw, and two free throws.

2. Describe a simulation of breaking a stick 1 m long into three pieces randomly. (Note that this means that the stick is broken in *two* places, each of which can be thought of as a coordinate between 0 and 1.) Run the simulation 20 times. Calculate the probability that the 3 "pieces" can be put together to form a triangle. (*Hint*: The 3 pieces will form a triangle only if the sum of any two of their lengths is greater than the third length.)

3. At a college dormitory, there are 5 public telephones that are used for outgoing calls. Suppose each telephone call lasts 3 min or less, and suppose you arrive at the telephones when they are all in use.

   a. Describe a simulation that will tell you the amount of time you will have to wait before a phone is free.

   b. Run the simulation 10 times. Find the average length of time you will wait for a telephone to become free.

4. In volleyball, one team serves the ball as long as that team keeps winning the volley that follows the serve. Suppose a certain team has a 50% probability of winning a volley while they are serving.

   a. Describe a simulation that will tell the number of consecutive serves the team will get before they lose the volley (and therefore the serve). Count the final losing volley as one of the serves in the sequence.

   b. Run your simulation 20 times. Find the average number of serves the team will get while they have the serve.

5. **a.** Describe a simulation for the following experiment: Choose 3 integers at random between 1 and 10, inclusive.

   b. Run your simulation 50 times. Calculate the probability that the three numbers selected will come up in increasing order. That is, if the numbers are a, b, and c, then $a \leq b \leq c$.

   c. Modify the simulation so that the numbers do not have to be integers. Run your simulation 50 times. Calculate the probability that these numbers are in increasing order.

Challenge Problems, ALGEBRA 2: EXPLORATIONS AND APPLICATIONS

Chapter 1 Challenge Set ·········

1. The population of a town since 1990 is modeled by $y = 10{,}240(1.05)^n$, where n is the number of years after 1990. (*Section 1.2*)

   a. For the years 1991 through 1995, find the change in population (to the nearest whole number) from the preceding year, and enter these figures in a table.

   b. Find an exponential model of the changes you entered in your table in part (a). How good a fit is the exponential model?

2. The area of the parallelogram $OPQR$ is $ad - bc$. (*Section 1.3*)

   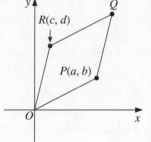

   a. Draw a new parallelogram formed by using the point $R(c + a, b + d)$ in place of $R(c, d)$, and completing the parallelogram similarly. Where does the point R now fall? What is the relationship between the area of the new parallelogram and the old one? (*Hint*: Area of a parallelogram = base × height.)

   b. Calculate the determinant of $Y = \begin{bmatrix} a & b \\ c + a & d + b \end{bmatrix}$. How is the determinant of

   $X = \begin{bmatrix} a & b \\ c & d \end{bmatrix}$ related to the determinant of Y? Does this relationship agree

   with the relationship between the areas of the two parallelograms that you found in part (a)?

3. A matrix of the form $\begin{bmatrix} a & b & c \\ 0 & d & e \\ 0 & 0 & f \end{bmatrix}$ is called a *triangular matrix*. Use this

   matrix and another of the same form (but using different letters) to show that the product of two triangular matrices is triangular. (*Section 1.4*)

★ 4. A matrix of the form $\begin{bmatrix} a & 0 & 0 \\ 0 & b & 0 \\ 0 & 0 & c \end{bmatrix}$ is called a *diagonal matrix*. If none of the

   numbers a, b, and c is 0, this matrix has a multiplicative inverse. Make a conjecture about the entries in the inverse, in terms of a, b, and c. Prove your conjecture by multiplying the two matrices in both of the possible orders. (*Section 1.4*)

5. *Extension* In a men's professional tennis match, the first person to win 3 sets wins the match. Suppose, in a match between Wang and Delgado, there is a $\frac{2}{3}$ probability that Delgado will win any given set. Describe a simulation of a match, and perform your simulation as many times as you can. From the trials of the simulation, calculate the experimental probability that Delgado will win any given match. Is your probability greater than, less than, or about equal to $\frac{2}{3}$?

Challenge Set 6 ···

In Exercises 1–6, a pair of corresponding values of a direct variation is given. Find the missing *x*-value or *y*-value that corresponds to the third value. (In Exercises 5 and 6, give your answer in terms of *a* and *b*.)

1. $y = 15$, when $x = 4$; $y = $ _?_ when $x = 10$ **2.** $y = 11.2$ when $x = 48$; $y = 49$ when $x = $ _?_

3. $y = \dfrac{9}{4}$ when $x = 12$; $y = \dfrac{27}{2}$ when $x = $ _?_   **4.** $y = 156$ when $x = 28$; $y = $ _?_ when $x = 35$

5. $y = 2a^2$ when $x = \dfrac{b}{3}$; $y = $ _?_ when $x = \dfrac{b^2}{a}$ **6.** $y = \dfrac{1}{3ab}$ when $x = 4a$; $y = \dfrac{1}{6a^2}$ when $x = $ _?_

7. In about 200 B. C. the Greek mathematician Eratosthenes estimated the circumference of Earth by knowing that the circular arc connecting the Egyptian cities Syene and Alexandria was about 5000 *stadia* in length and deducing that the central angle between these two cities was about 7.2°.

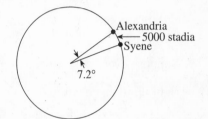

a. Assuming that arc length varies directly with central angle, calculate the circumference of Earth in stadia.

b. Assuming that 1 stadium = 0.168 km, convert your answer to part (a) to kilometers.

8. A container for a new sun-screen product is to be made in the shape shown, with height equal to twice its radius. The amount of plastic needed to make the container then varies directly with the *square* of the radius *r*. This means that the amount of plastic *A* is given by a function of the form $A = ar^2$, for some constant *a*. Suppose that a small size container with a radius of 1.1 in. requires 22.8 in.^{2} of plastic. How much plastic would be required for the large size, which has a radius of 1.8 in.?

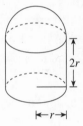

9. Suppose that the graph of a function has the property that whenever the point (x, y) is on the graph, so is the point (kx, ky) for any real number *k*.

a. Explain how you know that $(0, 0)$ must be on the graph of the function.

b. Show that if the equation of the function has the form $y = ax + b$, then *b* must be 0.

c. Suppose that the point $\left(\dfrac{2}{3}, 5\right)$ is on the graph of the function. Explain how you know that $(4, 8)$ is *not* on the graph. What point on the graph has *x*-coordinate 4?

Challenge Set 7 ···

Find k so that the line with the given equation has the given slope.

1. $3x - ky = 6$; slope $= 5$

2. $-2kx + 5y = 4$; slope $= -\dfrac{1}{2}$

3. $2x - 3ky = 7$; slope $= -6$

4. $\dfrac{2}{k}x - 8y = 12$; slope $= \dfrac{3}{4}$

Find an equation, in slope-intercept form, of the line with the same slope as the given line and with the given y-intercept.

5. $4x + 6y = 5$; y-intercept $= \dfrac{7}{3}$

6. $2x - 5y = -8$; y-intercept $= 6$

7. $ax - 4y = -3$; y-intercept $= -4$

8. $3x + 2by = 5$; y-intercept $= \dfrac{1}{2}$

9. Suppose every point on the graph of $y = -\dfrac{3}{2}x + 4$ is translated 2 units to the right. Find an equation of the new graph in slope-intercept form.

10. Suppose every point on the graph of $y = \dfrac{2}{3}x - 5$ is translated 4 units up and 6 units to the right. Find an equation of the new graph in slope-intercept form.

11. As Donya drives away from her house along a straight road, her position relative to an east-west and north-south coordinate system in miles is $(30t + 25, 40t)$, where t is the time, in hours, after her departure.

   a. After how long will her x-coordinate be 40? What will her y-coordinate be then?

   b. Find an equation, in slope intercept form, for her path in this coordinate system. (*Hint*: Let $x = 30t + 25$ and let $y = 40t$. Combine these equations into a single equation that does not involve t.)

12. a. As t increases from 0 to 1, the point $(8t - 2, 6t + 1)$ traverses a line segment. What are the endpoints of this segment?

   b. Find an equation, in slope-intercept form, of the line that contains the segment you found in part (a).

★ **13.** Suppose the graph of a function passes through the point $P(c, d)$ and suppose that if (x, y) is any other point on the graph, $y - d$ varies directly with $x - c$, with constant of variation k. Show that the graph is a line and find the slope and y–intercept of the line.

Challenge Set 8 ···

FOR USE WITH SECTION 2.3

Find each value of the function defined by $f(x) = -\frac{1}{4}x + 5$. (Give the values for Exercises 5–8 in terms of the given variables.)

1. $f(6)$

2. $f(-3)$

3. $f\left(-\frac{7}{2}\right)$

4. $f\left(\frac{4}{5}\right)$

5. $f(a + 3)$

6. $f(2c)$

7. $f\left(\frac{b}{3}\right)$

8. $f\left(k - \frac{1}{2}\right)$

The *composite* of two functions f and g, denoted $f(g(x))$, combines the two functions by using the output values of one as the input values of the other. For example, if $f(x) = 5x + 3$ and $g(x) = 2x$, then the composite function is $f(g(x)) = 5(2x) + 3 = 10x + 3$. For each pair of functions, find $f(g(x))$ and $g(f(x))$ and tell whether these two composite functions are equal.

9. $f(x) = 2x - 1;\ g(x) = x + 2$

10. $f(x) = x + 5;\ g(x) = x - 3$

11. $f(x) = \frac{1}{4}x + 5;\ g(x) = -3x + 4$

12. $f(x) = -\frac{2}{3}x + 1;\ g(x) = 6x$

13. *Writing* Suppose f and g are linear functions, as in Exercises 9–12. Even when $f(g(x)) \neq g(f(x))$, what common features do these two functions have? Are they both linear? Do their graphs have the same slope? If so, how is the common slope related to the slopes of f and g? If not, how do they differ?

Find the domain and range of each function.

14. $f(x) = |x|$

15. $f(x) = -\sqrt{x - 2}$

16. $f(x) = 3 - x^2$

Solve each equation or inequality.

17. $|x - 3| = 2$

18. $2|x + 4| = 11$

19. $\frac{1}{2}|3 - x| = 5$

20. $|x + 1| > 4$

21. $\frac{1}{3}|x - 5| \leq 1$

22. $|2x - 3| < 7$

For each of the following functions, find $f(a)$, $f(b)$, and $f(a + b)$ in terms of a, b, or both a and b.

23. $f(x) = x^2$

24. $f(x) = 2x + 5$

25. $f(x) = 3x$

26. $f(x) = \frac{1}{x}$

27. For which of the functions in Exercises 23–26 is it true that for *any* numbers a and b, $f(a + b) = f(a) + f(b)$? Make a conjecture based on your answer.

Challenge Set 9 ···

1. A computer service provider charges $10 for up to 5 hours of on-line time per month and $2 for every additional hour or part of an hour. (This means that, for example, if you spent 5.5 hours on line, you would be charged $12.)

   a. Graph the cost y of x hours of on-line time. Is y a function of x?

   b. Suppose the cost y of x hours of time were calculated according to the formula $y = 2x$. Graph this function on the same axes. How would the costs of the same amounts of time compare under the two rate schedules?

2. If (x_1, y_1), (x_2, y_2), (x_3, y_3), . . . , (x_n, y_n) are a set of data points, you can find the equation of the best-fit line as follows:

   Square all the x-coordinates and find the mean of these squares $\overline{x^2}$.
   Also, find the mean $\overline{x}$ of the x-coordinates.
   Let $s_x = \overline{x^2} - (\overline{x})^2$. The *slope* of the best-fit line is then given by

   $$a = \frac{\overline{xy} - \overline{x} \cdot \overline{y}}{s_x}$$

   where $\overline{y}$ is the mean of the y-coordinates and $\overline{xy}$ is the mean of all the products of the corresponding x- and y-coordinates. The best-fit line must also pass through $(\overline{x}, \overline{y})$. Use this procedure to find the equation of the best-fit line of the following data points: (3, 15), (5, 22), (6, 31), (7, 27), (9, 35).

3. The table below shows the number of utility patents issued by the U. S. Patent Office in decades since 1940.

Decade	1941–1950	1951–1960	1961–1970	1971–1980	1981–1990
Number of Patents (1000s)	308	430	585	687	737

   a. Draw a scatter plot of the data. Use a calculator to find the regression line of the data, using as the independent variable the number of decades after 1941–1950, and plot this line on the same axes.

   b. Most calculators will also find equations of other kinds of best-fit graphs, for example a *quadratic regression curve*. Find an equation for the quadratic regression curve or one of these other best-fit graphs and plot it on the same axes as the best-fit line you plotted in part (a). Is the other best-fit graph a better model for the data?

4. Find the exponential model of the data in Exercise 3 by finding the average growth factor, as you did in Chapter 1. If your calculator will find an equation of the *exponential regression curve*, graph this curve also. How does this curve compare with your models from parts (a) and (b)?

Challenge Set 10 ···

1. Suppose x, y, and z are three variables, any two of which can be compared by means of a scatter plot. For example, x might be a person's age, y might be the person's height, and z might be the same person's weight. The table below lists some of the possible correlations among the variables. Fill in the missing entry in each row.

Correlation between x and y	Correlation between y and z	Correlation between x and z
positive	negative	?
negative	?	negative
negative	negative	?
?	positive	positive

2. The two scatter plots at the right indicate how you can discover the formula for the correlation coefficient between two variables x and y. In both diagrams, the symbols $\bar{x}$ and $\bar{y}$ represent the mean of the x-coordinates and the mean of the y-coordinates, respectively.

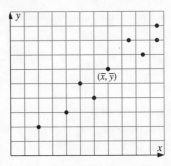 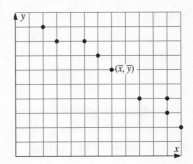

   a. Suppose (x, y) are the coordinates of one of the data points in either diagram. What can you say about the *sign* of the product $(x - \bar{x})(y - \bar{y})$ in each diagram? If two variables are strongly positively correlated, what would you expect to be true of the *sum* (and therefore also of the mean) of all such products? Answer the same question for variables that are strongly negatively correlated.

   b. Suppose (x_1, y_1) and (x_2, y_2) are the *only* two data points in a data set. Show that the *mean* of the two products mentioned in part (a) is

   $$\frac{x_1y_1 + x_2y_2}{2} - \frac{x_1 + x_2}{2} \cdot \frac{y_1 + y_2}{2}.$$

3. You can calculate the correlation coefficient between two variables x and y as follows: Let $s_x = \overline{x^2} - (\bar{x})^2$ and let $s_y = \overline{y^2} - (\bar{y})^2$. Then the correlation coefficient r of two variables x and y is given by

   $$r = \frac{\overline{xy} - \bar{x} \cdot \bar{y}}{\sqrt{s_x \cdot s_y}}.$$

   Use this formula to find the correlation coefficient of x and y with data points (3, 15), (5, 22), (6, 31), (7, 27), and (9, 35).

Challenge Set 11 ·····························

FOR USE WITH SECTION 2.6

In Exercises 1–4, a pair of parametric equations for the motion of an object is given, where *x* and *y* are in feet and *t* is in seconds. Find the speed of the object along its path.

EXAMPLE: $x = 5t + 3; y = 12t$

SOLUTION: Choose a time interval, say $0 \le t \le 1$. At $t = 0$ and $t = 1$, the object is at (3, 0) and (8, 12), respectively. Find the distance *d* between these two points.

$d = \sqrt{(8 - 3)^2 + (12 - 0)^2} = \sqrt{25 + 144} = \sqrt{169} = 13$. The point travels 13 ft in 1 s. Its speed is therefore 13 ft/s.

1. $x = 6t - 1; y = -8t + 5$

2. $x = \dfrac{3}{2}t; y = 2t + 3$

3. $x = \dfrac{7}{6}t + 1; y = 4t - 3$

4. $x = -4t - 1; y = \dfrac{21}{5}t$

Graph each pair of parametric equations for the given restriction on *t*.

5. $x = 2t$
$y = |t - 1|$
$0 \le t \le 2$

6. $x = t + 1$
$y = 2|t| - 1$
$-1 \le t \le 1$

7. $x = t - 2$
$y = t^2$
no restriction on *t*

An expression for *y* as a function of *x* and one of a pair of parametric equations are given. Find the other parametric equation so that the two parametric equations combine to express *y* in terms of *x* as specified.

8. $y = \dfrac{5}{2}x + 1$
$x = -2t$

9. $y = 8x + 20$
$y = 2t + 6$

10. $y = -3x + 17$
$x = 6t + 1$

11. *Writing* Use the parametric equations $x = 6t$ and $y = -3t + 5$ to express *y* in terms of *x*. Then use the reverse expressions $x = -3t + 5$ and $y = 6t$ to do the same. Graph both equations on one set of axes. What relationships do you notice involving slopes and *x*- and *y*-intercepts? How are the two equations related algebraically?

12. At noon, a car leaves Centerville heading east at 40 mi/h. At the same time a truck leaves Southport heading north at 30 mi/h. Southport is 70 mi south of Centerville, as shown.

a. Write parametric equations for the distance of each vehicle from Centerville as functions of *t*, the time after noon.

b. Determine how far apart the two vehicles are at 1:30 P.M.

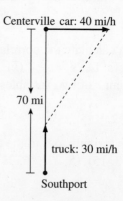

Centerville car: 40 mi/h

70 mi

truck: 30 mi/h

Southport

Chapter 2 Challenge Set ·····························

FOR USE AFTER CHAPTER 2

1. Each algebraic description below defines a *family* of lines. Graph several of the lines in the family and give a *geometric* characterization of the family. (*Sections 2.1 and 2.2*)

   a. Lines with equations of the form $y = 2x + k$ for all possible real values of k.

   b. Lines with equations of the form $y - 3 = a(x + 2)$ for all possible real values of a.

   c. Lines with equations of the form $y - 5 = -\frac{2}{3}(x - h)$, for all possible real values of h.

2. a. For the following data points, find the mean $\bar{x}$ of the x-coordinates and the mean $\bar{y}$ of the y-coordinates: (5, 7), (6, 11), (8, 17), (9, 15), (12, 20)

   b. Using a calculator, find the equation of the best-fit line for the data points in part (a). Does the point $(\bar{x}, \bar{y})$ lie on the line?

   c. Make up another set of data points and find the best-fit line. Does the point $(\bar{x}, \bar{y})$ lie on the line? (*Section 2.4*)

3. a. Suppose there is a strong positive correlation between variables x and y, and suppose there is a weak correlation between variables y and z. Describe the correlation between x and z. Give an example of the situation using real-world variables.

   b. Suppose there is a strong positive correlation between variables x and z. Is it possible that there is a variable y such that x and y are weakly correlated, and y and z are also weakly correlated? If so, give an example of such a relationship using real-world variables. If not, explain why not. (*Section 2.5*)

4. Suppose a line has parametric equations $x = g(t)$, $y = h(t)$, and suppose the point (p, q) is on the line. Find an expression for the *slope* of the line using g, h, p, and q and a value of t, t_0, such that $g(t_0) \neq p$. (*Section 2.6*)

5. *Extension* Work in a group of 2 or more students. Start by making up the equation of a line, in slope-intercept form, such as $y = \frac{1}{2}x + 1$. Each student in the group should make up at least 6 data points, *none of which lies on the line*, so that the equation for the best-fit line for the points comes as close to the given equation as possible. After each student in the group has made up 6 points, combine all the points and find *one* best-fit line. Is this line closer to having the given equation than any of the individual lines? If the slope of the calculated best-fit line is too large, how could you change one or more of the points to produce a smaller slope? How could you change one or more points to produce a smaller y-intercept?

Challenge Set 12 ···

Solve each equation.

1. $25(3^x) = 225$

2. $0.4(2^x) = 12.8$

3. $\frac{1}{4}(10^x) = 2500$

4. $\frac{1}{16}(0.2^x) = 0.0005$

5. $0.25(4^x) = 256$

6. $1000\left(\frac{1}{4}\right)^x = 62.5$

7. A cube is cut in half, each of the two pieces created is cut in half, each of the four pieces created is cut in half, and so on. (The first three cuts are shown as a thick line, a thin line, and a dotted line in the diagram.)

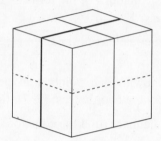

a. Suppose an edge of the original cube is 12 in. long. Complete the table below for the stages in the process, dimensions of each piece, and volume of each piece.

Stage	Length	Width	Height	Volume
0	12	12	12	1728
1	12	6	12	864
2	6	6	12	432
3	?	?	?	?
4	?	?	?	?
5	?	?	?	?

b. Write an exponential decay equation for the volume as a function of the number of stages in the process.

c. Rewrite the function you wrote in part (b) using a as the length of an edge of the original cube.

8. The "Cantor Middle-Third Set" is a set formed in stages as follows:

Stage 0: The unit interval.

Stage 1: The interval is cut into three equal pieces, and the middle third is dropped out.

Stage 2: Each of the pieces left from the last stage is cut into three equal pieces, and the two middle pieces are dropped out.

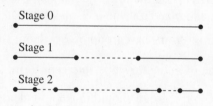

a. How many pieces of the interval remain after the nth stage of the process. Give your answer in terms of n.

b. Write an exponential function that gives the total length of the remaining pieces of the unit interval after the nth stage.

Challenge Set 13 ··

Simplify using the properties of exponents.

1. $[2^{1/6} \cdot 8^{1/2}]^3$

2. $\dfrac{625^{1/3}}{(25^{1/3})^{1/2}}$

3. $27^{1/2} \cdot 3^{-1} \cdot 9^{1/4}$

4. $(5^{1/6} \cdot 125^{1/6})^3$

5. $\dfrac{4^{1/3} \cdot 16^{2/3}}{2^{1/3}}$

6. $[(0.01)^{-1/6} \cdot 100^{1/3}]^2$

Solve each equation.

Example: $\frac{1}{3}x^{4/3} = 27$

Solution: $\frac{1}{3}x^{4/3} = 27 \quad \Rightarrow \quad x^{4/3} = 81 \quad \Rightarrow \quad (x^{4/3})^{3/4} = 81^{3/4} \quad \Rightarrow$

$x = (81^{1/4})^3 = 3^3 = 27$

7. $2x^{3/2} = 16$

8. $125x^{5/4} = \dfrac{1}{25}$

9. $\frac{1}{9}x^{-4/3} = 9$

Example: $2^{3x+1} = 4$

Solution: $2^{3x+1} = 4 \quad \Rightarrow \quad 2^{3x+1} = 2^2 \qquad$ Equating exponents:

$3x + 1 = 2 \quad \Rightarrow \quad x = \dfrac{1}{3}$

10. $27^{x+2} = 9^x$

11. $5 \cdot 5^{4x-2} = 25^{x+3}$

12. $4^{2x+1} = 8^{x-1}$

13. A bacteria culture has 135,000 bacteria at 12:00 noon and increases by 5% every 4 hours.

   a. By what factor does the culture increase every 4 hours? If t represents the number of hours since 12:00 noon, what expression gives the number of 4-hour periods since 12:00 noon?

   b. Write an equation for the size y of the culture as a function of the time t after 12:00 noon.

   c. Find the size of the culture at 3:00 P.M.

14. *Writing* Your calculator probably has no trouble evaluating the expression $(-4)^\wedge 5$, which means $(-4)^5$, and it may even evaluate $(-4)^\wedge(1/5)$. But if you enter the expression $(-4)^\wedge(3/5)$, meaning $(-4)^{3/5}$, you will probably get an error message. Why do you think this happens? (*Hint*: $\dfrac{3}{5} = \dfrac{6}{10}$.)

Challenge Set 14 ···

Rewrite each function in the form $y = a \cdot 2^{x/d}$.

1. $y = 150(1.06)^x$ **2.** $y = 1200(1.08)^x$ **3.** $y = 4.6(1.14)^x$

Rewrite each function in the form $y = a \cdot \left(\dfrac{1}{2}\right)^{x/h}$.

4. $y = 72.5(0.8)^x$ **5.** $y = 1450(0.92)^x$ **6.** $y = 128(0.65)^x$

Use the graph of the corresponding function of the form $y = ab^x$ to solve the given equation.

7. $15 = 2(1.4)^x$ **8.** $128 = 10(1.7)^x$ **9.** $24{,}000 = 1500(2.6)^x$

10. $81 = 108(0.85)^x$ **11.** $15 = 75(0.63)^x$ **12.** $42 = 70\left(\dfrac{7}{8}\right)^x$

Solve each inequality by graphing the corresponding function $y = ab^x$.

13. $7.6(2.4)^x > 135$ **14.** $140(0.8)^x \leq 56$ **15.** $3.5(1.3)^x > 39$

16. For each of the functions below, describe how the graph of $y = 2(1.5)^x$ is altered to produce the graph of the given function.

    a. $y = 2(1.5)^x + 4$ **b.** $y = 2(1.5)^{x-3}$ **c.** $y = 2(1.5)^{-x}$

17. The population of a town was 25,000 in 1975 and had doubled by 1987. Assuming exponential growth, write an equation of the form $y = ab^x$ for the population y of the town x years after 1975.

18. Marvin Williams bought an antique chair for $250 at the beginning of 1990 and sold it for $350 in July of 1994, about 3.5 years later.

    a. Marvin could have invested the $250 instead of buying the chair. Suppose he had invested the money at $r\%$. Write an equation for the value of his investment y after 3.5 years, as a function of r.

    b. What interest rate would have given Marvin the same profit on his investment after 3.5 years as he made on the sale of the chair?

Challenge Set 15 ···

1. a. Suppose a quantity y is determined by the function $y = e^t$, where t represents time. Find the doubling time for y to 3 decimal places.

b. Rewrite the function $y = e^t$ in the form $y = \left(\frac{1}{2}\right)^{kt}$.

2. Use the logistic function $y = \frac{a}{1 + be^{-t}}$.

a. What number does y get close to as t approaches $-\infty$, that is, as t takes on negative values that are larger and larger in absolute value? Explain.

b. What number does e^{-t} approach as t approaches $+\infty$? What number, in terms of a and b does y approach as t approaches $+\infty$, that is, as t gets larger and larger? Explain.

★ **3.** You have seen that the expression $\left(1 + \frac{r}{n}\right)^n$ is a good approximation of e^r

when n is very large. To see an algebraic proof of why this should be true,

let $u = \frac{n}{r}$.

a. Write n and $\frac{r}{n}$ in terms of r and u only. What happens to u as n gets large?

b. Write the expression $\left(1 + \frac{r}{n}\right)^n$ in terms of r and u only, and explain why it

is a good approximation of e^r.

4. The number e can also be approximated by sums of the form

$$1 + \frac{1}{1!} + \frac{1}{2!} + \frac{1}{3!} + \ldots + \frac{1}{n!}$$

where $n!$ represents the product $1 \cdot 2 \cdot 3 \cdot \ldots \cdot n$.

a. Approximate e by using the first 5 terms of this series.

b. Which of the following do you think is approximated by

$$1 + \frac{2^1}{1!} + \frac{2^2}{2!} + \frac{2^3}{3!} + \ldots + \frac{2^n}{n!} : e + 2, 2e, e^2, \text{ or } \frac{1}{2}e?$$

Determine an approximate value for $1 + \frac{a^1}{1!} + \frac{a^2}{2!} + \frac{a^3}{3!} + \ldots + \frac{a^n}{n!}$.

Challenge Set 16 ···

FOR USE WITH SECTION 3.5

1. When the shock absorbers of a car are worn out and you push down on one fender, the car undergoes *underdamped vibration*, that is, it bounces up and down in progressively smaller bounces. Suppose $y = 0$ represents the car's equilibrium (rest) height, and suppose after you push down on the fender, the car bounces up to a maximum height of 0.65 in. and down to a minimum height of –0.32 in.

   a. Write an exponential function of the form $y = ab^n$ for the height of the car above (or below) its equilibrium height on the nth bounce (up or down). (*Hint:* A negative number raised to integer powers will be alternately positive and negative, depending on whether the integer is odd or even.)

   b. After how many bounces will the car be within 0.1 in. of equilibrium?

2. *Writing* Describe the behavior of a function of the form $y = ab^{-x} + c$, where a, b, and c are positive numbers, as x gets very large. Give an example of a real-world phenomenon that could be modeled by an equation of this form.

3. How many times do the graphs of $y = x^2$ and $y = 2^x$ intersect? Answer this question by using a graphing calculator to graph the two functions on one set of axes. What are the coordinates of the intersection points to the nearest hundredth?

4. Suppose $y = ab^2$ and $y = ac^2$ are exponential functions for which $a \neq 0$ and $b \neq c$.

   a. By trying some examples for various values of a, b, and c, determine how many times the graphs of the two functions can intersect. What seem to be the coordinates of the intersection point(s) in terms of a, b, and c?

   ★ b. Prove your answer to part (a) algebraically.

5. a. Suppose you pick a point (r, b^r) on the graph of the exponential function $y = b^x$ and draw the segment connecting this point to the point $(0, 1)$, which is also on the graph. Find the equation of this line in slope-intercept form, in terms of r and b.

   ★ b. The midpoint of the segment connecting $(0, 0)$ and $(r, 0)$ on the x-axis is $\left(\frac{r}{2}, 0\right)$. Show that the y-coordinate of point A directly above this point and on the line in part (a) is $\frac{b^r + 1}{2}$. What is the y-coordinate of the corresponding point B on the graph of $y = b^x$?

   ★ c. Show that the y-coordinate of point A in part (b) is larger than the y-coordinate of the corresponding point B using the following method: Expand the expression $\frac{1}{2}(b^{r/2} - 1)^2$ and use the fact that this expression must always be greater than or equal to 0.

Chapter 3 Challenge Set ·······························

FOR USE AFTER CHAPTER 3

1. Suppose a large number N of dice is rolled. All the dice that come to rest showing 6 are discarded. The remaining dice are rolled again, and the process is repeated.

   a. What fraction of the dice would you expect to remain after the first roll?

   b. Write an equation that gives y, the expected number of dice remaining after n rolls, as a function of n. About how many rolls would it take before the number of dice remaining is less than $\frac{1}{5}N$? *(Section 3.1)*

2. A computer shows the graph of an exponential equation, such as $y = 2^x$, as a solid curve. In particular, this function has a value when x is an *irrational* number, such as $\sqrt{3} = 1.73205 \ldots$. *(Section 3.2)*

   a. Assuming that $2^{\sqrt{3}}$ is defined, which of the following numbers do you think are less than $2^{\sqrt{3}}$: $2^{1.7}$, $2^{1.73}$, $2^{1.74}$, $2^{1.7321}$, $2^{1.732}$? Explain how you know that each of these numbers has a well-defined value.

   b. Simplify each of the following expressions: $(2^{\sqrt{3}})^{\sqrt{3}}$; $2^{1 + \sqrt{3}} \cdot 2^{1 - \sqrt{3}}$

3. If you hold a string at two points and let the length of the string between these two points hang loose, the string assumes the shape of a *catenary*. This is a curve that has an equation of the form $y = a(e^{bx} + e^{-bx})$ for some constants a and b. *(Section 3.4)*

   a. Suppose that the coordinate system for a catenary has inches as its units and suppose the point $(0, 1)$ is on the graph of the catenary. Find the value of a.

   b. Suppose the same catenary as in part (a) passes through the point $(1, 5)$. Use a graphing calculator to find the value of b to the nearest tenth of an inch. (*Hint*: Graph $y = a(e^x + e^{-x})$.)

4. **Extension** If you borrow money to buy a house or a car, the loan is usually *amortized*. This means that you repay the money in equal installments, usually monthly. If you borrow a total amount P at a monthly rate r and you repay the loan over the course of n months, then your monthly payment M is given by the formula

$$M = \frac{rP}{1 - (1 + r)^{-n}}.$$

   Working with a partner, make up a rate r, a principal P, and a monthly payment M and give these data to your partner, who will calculate the number of months it will take to repay the loan. (*Hint*: Use a graph to do this, after solving the equation above for $(1 + r)^n$.) Do this calculation for several different values of r, P, and M.

Challenge Set 17 ··

1. a. Find the inverse of the linear function $f(x) = ax + b$ in terms of a and b. How is the slope of a linear function $f(x)$ related to the slope of its inverse function $f^{-1}(x)$?

   b. Find the composite functions $f(f^{-1}(x))$ and $f^{-1}(f(x))$.

2. Suppose the point $(1, 4)$ is on the graph of $y = g(x)$, and the point $(3, -1)$ is on the graph of $y = g^{-1}(x)$. Find an equation for g and an equation for g^{-1}.

3. a. In order for a function to have an inverse, it must be one-to-one. This means that any two points on the graph with different x-coordinates cannot have the same y-coordinate. Describe a geometric property of the graph, as it relates to straight lines, that is equivalent to this condition.

   b. Explain, using the definition of inverse functions, why the condition stated in part (a) must hold for a function that has an inverse.

Graph each function. Use the condition described in Exercise 3 to decide whether the function has an inverse. If it does, find an equation for the inverse function and graph it on the same axes.

4. $f(x) = \dfrac{1}{x} + 2$ **5.** $g(x) = 2\left|x\right|$ **6.** $f(x) = x^3$

7. $f(x) = -x^2$ **8.** $h(x) = \dfrac{4}{x - 3}$ **9.** $g(x) = x + \left|x\right| + 1$

10. Suppose a function f has the following property:

   For any x-values a and b with $b > a$, $f(b) > f(a)$.

   a. Sketch the graph of two functions, one linear and one nonlinear, that have this property. What name would you give to a function with this property?

   b. Explain how you know that a function with the property above has an inverse function.

   c. Suppose the last ">" sign in the property above is changed to "<." Answer parts (a) and (b) for the new property produced by this change.

★ **11. a.** Sketch the graph of the function $f(x) = x\left|x\right|$. Explain how you know that this function has an inverse function.

   b. Find an equation for the inverse of the function you graphed in part (a). (*Hint*: You may have to define the function *piecewise*: that is, $f(x) = \ldots$, if $x \geq 0$; $f(x) = \ldots$, if $x < 0$. To help find the inverse function, write down some pairs (x, y) on the graph.)

Challenge Set 18 ·······································

Solve by first converting the given equation to exponential form.

1. $\log_3 x = 4$

2. $\log 0.001 = x$

3. $\log_{1/2} 16 = x$

4. $\log_x 25 = \dfrac{1}{2}$

5. $\log_x 8 = \dfrac{3}{2}$

6. $\log_x \dfrac{1}{64} = 6$

7. $\log_{32} 8 = x$

8. $\log_8 x = -\dfrac{2}{3}$

9. $\log_x 5 = 2$

EXAMPLE: $\log_{3x} 36 = 2$

SOLUTION: $\log_{3x} 36 = 2 \Rightarrow (3x)^2 = 36 \Rightarrow 9x^2 = 36 \Rightarrow x^2 = 4 \Rightarrow$

$x = 2$, since a base cannot be negative.

10. $\log_3 (x^2 + 2) = 3$

11. $\log_2 32 = x - 1$

12. $\log_{1/5} (x^2 + 4) = -3$

13. The decibel level of a sound is a measure of its loudness. The decibel level d of a given sound is defined as

$$d = 10 \log \left(\frac{I}{I_0} \right),$$

where I is the intensity of the given sound and I_0 is the intensity of a barely audible sound.

a. A conversation at normal levels has an intensity of about $10^6 \cdot I_0$. What is the decibel level of such a conversation?

b. A jet taking off has a decibel level of 120. Find the intensity of the sound in terms of I_0.

14. You know that you can approximate e^x by evaluating the expression

$\left(1 + \dfrac{x}{n} \right)^n$ for large values of n. That is, $e^x \approx \left(1 + \dfrac{x}{n} \right)^n$ when n is large.

a. Substitute $\ln a$ for x in the approximation above. Simplify e^x for this value of x and rewrite the approximation.

b. Solve the approximation for $\ln a$ in terms of a and n. Use it to find $\ln 3$. (*Hint*: Use $n = 64$).

Challenge Set 19 ·····························

FOR USE WITH SECTION 4.3

Write each expression in terms of $\log_3 a$, $\log_3 b$, and $\log_3 c$.

1. $\log_3 \dfrac{(ab^5)^{1/2}}{c^3}$

2. $\log_3 [(a^4b) \cdot c^{-4}]^{1/5}$

3. $\log_3 \dfrac{b^{3/2}}{(a^2bc^3)^{1/2}}$

Given that $2^a = 3$ and $2^b = 5$, find each logarithm in terms of a, b, or both.

4. $\log_2 15$

5. $\log_2 9$

6. $\log_2 0.6$

7. $\log_2 \dfrac{1}{125}$

8. $\log_2 75$

9. $\log_2 5^7$

10. $\log_2 10$

11. $\log_2 12$

Use each equation to express y as a function of x.

12. $\log_7 y = -3 \log_7 x$

13. $\log_3 2x + \log_3 y = 2$

14. $2 \log_n x - \log_n 2y = 0$

15. $4 \log_5 x + \log_5 y = 1$

16. Explain how you know that $\log_n x$ is a one-to-one function.

★ **17.** Prove that, for any positive numbers a and b,

$$a^{\log b} = b^{\log a}.$$

(*Hint*: Take the log of each side separately. Then use the properties of logarithms and the result of Exercise 16.)

18. Let the number T_n be defined by the equation

$$T_n = \log \frac{1}{2} + \log \frac{2}{3} + \log \frac{3}{4} + \ldots + \log \left(\frac{n}{n+1} \right).$$

a. Use the properties of logarithms to simplify the expression for T_n.

b. What happens to the number T_n as n gets larger and larger?

Challenge Set 20 ···

Graph each equation and tell how the graph is related to the graph of
$y = \log_2 x$.

1. $y = 5 \log_2 x$ **2.** $y = 3 + \log_2 x$ **3.** $y = \log_2 (x - 5)$

Solve each inequality.

4. $3 \log_8 x < 2$ **5.** $\log_2 (x - 4) \geq 3$ **6.** $\log_3 (x + 4) - \log_3 5 < 3$

7. Suppose M, b, c, and d are positive numbers and none of the numbers b, c, or d equals 1. Explain how you know that

$$\frac{\log_c M}{\log_c b} = \frac{\log_d M}{\log_d b}.$$

Solve each equation.

8. $\log_6 x + \log_6 (x - 5) = 2$ **9.** $\log_5 (x + 1) + \log_5 (x - 3) = 1$

10. $3^{2x} - 10 \cdot 3^x + 9 = 0$ **11.** $(\log_2 x)^2 - \log_2 x - 12 = 0$

12. $\log_8 (\log_2 x) = 1$ **13.** $\log_2 (\log_2 16) = \log_7 x$

14. $2^{x + 2} = 3^x$ **15.** $7^{2x} = 3^{x + 1}$

16. Use the change-of-base formula to solve the equation

$$(\log_3 x)(\log_4 3) = 2.$$

17. In 1990, Toni invested \$1500 at an annual rate of 6%, compounded monthly. Her friend Maya invested \$1550 at an annual rate of 4.8%, also compounded monthly.

a. Set up an equation that states that the two investments have the same value after n months.

b. After how many months will the two investments have the same value?

Challenge Set 21

Suppose a set of data is modeled by each equation. Find an equation relating y and x.

1. $\log_2 y = \frac{1}{3} \log_2 x + 3$

2. $\log_4 y = 4x + \frac{5}{2}$

3. $\ln y = 3x + \ln 15$

4. $\log_6 y = 2 \log_6 x + 1$

The working-age population of a town and the numbers of full-time jobs available in the town are given for the years 1990–1995 in the table below. Use this table in Exercises 5–8.

Year	Working-age Population	Number of Full-time Jobs
1990	8125	10,650
1991	8780	10,740
1992	9475	10,950
1993	10,240	11,270
1994	11,040	11,680
1995	11,950	12,290

5. Find an exponential model for the working-age population as a function of the number of years after 1990. Find a power function that gives (working-age population – 8125) as a function of the same variable. Which model is a better approximation of the data?

6. Find an exponential model for the number of full-time jobs as a function of the number of years after 1990. Find a power function that gives (number of full-time jobs – 10,650) as a function of the same variable. Which model is a better approximation of the data?

7. Graph the two better models that you found in Exercises 5 and 6. (*Hint*: Don't forget to add the constant you subtracted in the case of the power functions.)

8. When will the working-age population of the town surpass the number of full-time jobs? Will the number of full-time jobs again overtake the population at some later date?

9. *Writing* On the same coordinate system, have your calculator graph the functions $y = \log x$ and $y = x^{1/n}$ for a value of n between 5 and 10. What appears to be true about the relative sizes of corresponding function values of the two functions? Now use your calculator to find the values of these functions for very large values of x, for example, $x = 10^{20}$. Do the relative sizes of the y-values agree with your observation of the two graphs? Why or why not?

Chapter 4 Challenge Set ·······························

FOR USE AFTER CHAPTER 4

1. A local income tax rate schedule is as follows: wage earners pay 3% of earned income up to $40,000, and 5% of any amount in excess of $40,000. (*Section 4.1*)

   a. Graph the function that relates income x with the amount of tax paid y.

   b. Explain how you know that the function you graphed in part (a) has an inverse. Give an algebraic definition of the inverse function, using a piecewise definition, that is, use a definition of the form $f(x) = \ldots$ if $x \le a, f(x) = \ldots$ if $x > a$.

2. Prove that, for any positive numbers a and b,
$$\log_a b \cdot \log_b a = 1.$$

   (*Hint*: Let $x = \log_a b$. Convert to exponential form. Rewrite the equation with base b.) (*Section 4.2*)

3. For $a > 1$, the quantity $\ln a$ can be represented geometrically as the area under the graph of $y = \dfrac{1}{x}$ between 1 and a. For each pair of values, express the area under the graph of $y = \dfrac{1}{x}$ between the two values as the natural logarithm of a single quantity. (*Section 4.3*)

   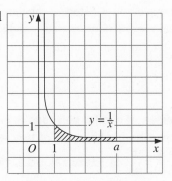

   a. $x = a, x = b$ **b.** $x = a, x = ab$ **c.** $x = \dfrac{b}{a}, x = b$

★ 4. **a.** Use the change-of-base formula to prove that if $x > 0$, $b > 0$, and $b \ne 1$, then
$$\log_b x = 2 \log_{b^2} x.$$

   b. Make a conjecture about the equation in part (a). Prove your conjecture.

   (*Section 4.4*)

5. *Extension* The function $\ln (1 + x)$ can be approximated by the formula
$$\ln (1 + x) = x - \frac{x^2}{2} + \frac{x^3}{3} - \frac{x^4}{4} + \ldots \; .$$

   Use this formula to show that $\ln (1 + x)^2 = 2 \ln (1 + x)$ by the following method. Rewrite $(1 + x)^2$ as $1 + (2x + x^2)$ and evaluate the right-hand side of the equation above with x replaced by $(2x + x^2)$. Carry out the evaluation through three terms. Expand your result, and compare the first three terms with the first three terms of $2 \ln (1 + x)$.

Challenge Set 22 ··

FOR USE WITH SECTION 5.1

Solve each equation. Give solutions to the nearest tenth when necessary.

1. $\frac{1}{5}(x-3)^2 = 45$　　　**2.** $\frac{1}{18x^2 - 4} = 2$　　　**3.** $\frac{1}{3x^2} + 2 = 5$

4. $8 - \frac{3}{x^2} = 6$　　　**5.** $\frac{3}{(x+4)^2} = 5$　　　**6.** $\frac{2x}{x-2} = -x$

★ **7.** Suppose $y = ax^2$, with $a \neq 0$, is the equation of a parabola. Show that if $p \neq 0$ and $q \neq 0$, then the parabola cannot contain *both* of the points (p, q) and $(2p, 2q)$. (*Hint*: Suppose both of these points were on the parabola. Find two equations that would be satisfied by p and q. Divide corresponding sides of the two equations.)

8. Suppose the square-root key on your calculator is broken and you want to solve the equation $x^2 = 10$. The following algorithm will find a solution: Start with a guess, x_1. In this case, use $x_1 = 3$. Divide 10 by x_1, average the result with x_1, and call the average x_2. Repeat the steps in the last sentence, using x_2 in place of x_1, calling the new result x_3, and so on. Stop when the numbers x_n and x_{n+1} are the same on your calculator. Use this method to solve each equation.

　a. $x^2 = 15$　　　**b.** $x^2 = 37$　　　**c.** $5x^2 = 29$　　　**d.** $10x^2 = 3$

Recall that the ratio of the areas of similar polygons is the square of their scale factor. For example, the ratio of the areas of two regular pentagons with sides of length 3 and 5 is $\left(\frac{3}{5}\right)^2 = \frac{9}{25}$. In Exercises 9-12, two regular polygons have sides of lengths s_1 and s_2 and areas A_1 and A_2. Find the missing side length. Give your answer to the nearest hundredth.

　9. $A_1 = 11$, $A_2 = 5$, $s_1 = 14$, $s_2 = ?$　　　**10.** $A_1 = 17$, $A_2 = 28$, $s_1 = ?$, $s_2 = 9.2$

11. $A_1 = 54$, $A_2 = 31$, $s_1 = ?$, $s_2 = 6.5$　　　**12.** $A_1 = 10.5$, $A_2 = 8$, $s_1 = 15$, $s_2 = ?$

Challenge Problems, ALGEBRA 2: EXPLORATIONS AND APPLICATIONS

Challenge Set 23 ···

FOR USE WITH SECTION 5.2

In Exercises 1-6, two points on the graph of $y = a(x - h)^2 + k$ are given, as is the value of h. Find the values of a and k and then write the equation.

1. $(2, 7)$, $(-1, 16)$; $h = 1$

2. $(-2, 4)$, $(1, -26)$; $h = -3$

3. $(6, -1)$, $(0, 5)$; $h = 4$

4. $(1, -7)$, $(4, -25)$; $h = -2$

5. $(-8, 13)$, $(-4, 1)$; $h = -5$

6. $(-2, 19)$, $(1, 1)$; $h = \dfrac{1}{2}$

Find the new equation when the graph of the given equation undergoes the given transformation.

7. The graph of $y = 4(x - 5)^2 - 1$ is shifted 3 units up and 4 units to the left.

8. The graph of $y = \dfrac{1}{2}(x + 2)^2 - 3$ is shifted 2 units down and 5 units to the right.

9. The graph of $y = 2(x - a)^2 + b$ is shifted c units down and d units to the left.

In Exercises 10 and 11, find the equation, in vertex form, of the parabola that satisfies the given conditions.

10. The parabola has its maximum on the line $y = 3$, is symmetric about the line $x = 4$, and contains the point $(1, 0)$.

11. The parabola has its minimum on the line $y = -2$ and passes through the points $(7, 2)$ and $(-1, 2)$.

★ **12.** By solving the equation $y = a(x - h)^2 + k$ for x, in terms of the other variables, show that if two points on the parabola with this equation have the same y-coordinate, they are equidistant from the line of symmetry of the parabola. Assume $a \neq 0$.

★ **13.** At a certain "height" y, the parabola whose equation is $y = a(x - h)^2 + k$, with $a > 0$, has width $\dfrac{1}{a}$. Find this value of y in terms of a, h, and/or k.

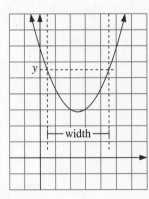

Challenge Set 24 ···

FOR USE WITH SECTION 5.3

Find the equation, in intercept form, of the parabola passing through the three given points.

1. $(1, 0), (-3, 0), (-2, -6)$

2. $(-2, 0), (-4, 0), (-1, 9)$

3. $(2, 0), (6, 0), (4, 20)$

4. $(-1, 0), (5, 0), (3, 18)$

5. $(3, 0), (7, 0), (5, 1)$

6. $\left(\frac{1}{2}, 0\right), \left(-\frac{5}{2}, 0\right), (2, -27)$

Solve each quadratic inequality.

Example: $2(x - 1)(x + 3) > 0$

Solution: Draw the graph of $y = 2(x - 1)(x + 3)$. When $x < -3$, the graph is above the x-axis, so $y > 0$ for these x-values. The graph is also above the x-axis when $x > 1$. Therefore, the solution consists of all real numbers x such that $x < -3$ or $x > 1$.

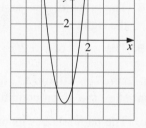

7. $3(x - 2)(x + 4) \le 0$

8. $-\frac{1}{2}(x - 3)(x - 1) < 0$

9. $-\frac{3}{4}(x + 7)(x + 3) \ge 0$

10. $2(x + 2)(x - 5) > 0$

Write each function in intercept form and find the x-intercepts of the graph.

11. $y = 2x^2 - 6 - 8$

12. $y = (3x - 6)\left(\frac{1}{2}x - \frac{5}{2}\right)$

13. $y = 21 + 4x - x^2$

★ **14. a.** Use intercept form to explain how you know that if a quadratic function has the form $y = x^2 + bx + c$ and it has two x-intercepts, then c is the product of these two intercepts.

b. How can you modify the statement in part (a) to take into account the general quadratic function $y = ax^2 + bx + c$, where $a \ne 0$, with two x-intercepts?

c. Suppose one solution of the equation $4x^2 + bx + 3 = 0$ is $\frac{1}{2}$. Find the other solution.

Challenge Set 25 ···

1. **a.** Find, in terms of a, the x-coordinate of the vertex of the parabola whose equation is $y = ax^2 + x$. Substitute your value into the equation to get the y-coordinate.

   b. Graph several of the parabolas described in part (a) for various values of a. Mark the vertex of each parabola. Make a conjecture about the curve described by the vertices.

   c. Prove your conjecture by finding an equation that relates each y-coordinate of a vertex with the corresponding x-coordinate.

2. An equation of the form $(x - h)^2 + (y - k)^2 = r^2$, where h, k, and r are constants, defines a circle in the coordinate plane with center (h, k). Find the center of each circle by putting the given equation in the form above.

   a. $x^2 + y^2 - 24x + 10y = 0$ **b.** $x^2 + y^2 - x - 3y - 3 = 0$

3. Write the equation $y = (x - r)(x - s)$ in vertex form to show that the x-coordinate of the vertex is the average of the x-intercepts of the graph.

4. **a.** By completing the square with an x^2 term, rewrite the expression $x^4 + 64$ as a difference of two expressions each of which is a perfect square.

   b. Factor the expression $x^4 + 64$ by factoring the expression you found in part (a).

★ 5. The curator of a nature preserve needs to tranquilize an ailing monkey so that it can be treated and released. Just as the curator, aiming directly at the monkey, shoots the tranquilizer dart, the monkey jumps from the tree where it had been sitting.

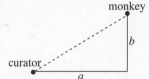

   a. Suppose the path of the dart is given by the parametric equations

   $$x = v_x t; \; y = -16t^2 + v_y t$$

   and the monkey's height h is given by $h = b - 16t^2$, where time t is measured from the moment the dart is released and the monkey jumps. Solve the first of these equations for t in terms of x and v_x and use this equation to eliminate t in the other two equations.

   b. Aiming the dart at the monkey means that $\dfrac{v_x}{v_y} = \dfrac{b}{a}$. Show that if this condition holds, the dart will hit the monkey. (*Hint:* How high is each when $x = a$?)

Challenge Set 26 ···

Use the quadratic formula to find the solutions of each equation.

1. $3x^2 - \sqrt{2}x - 5 = 0$

2. $\sqrt{5}x^2 + 6x - 3\sqrt{5} = 0$

3. $\sqrt{3}x^2 - 10\sqrt{2}x + 2\sqrt{3} = 0$

Find _k_ so that each quadratic equation has one solution.

4. $x^2 - \dfrac{3}{2}x + k = 0$

5. $2kx^2 + 8x + 6 = 0$

6. $3x^2 - 2kx + 4 = 0$

7. Lionel Washington invested $5000 two years ago, and then one year ago he invested another $2500 in the same account. He now has $8268 in the account. His investment has been drawing interest at the same annual rate, compounded annually, for both years.

    a. Let r = the annual rate of interest on his investments. Write expressions for the current value of his $5000 investment and of his $2500 investment, in terms of r.

    b. Write and solve a quadratic equation to find the rate of interest his investments have earned.

8. a. Write the two solutions of the equation $ax^2 + bx + c = 0$ separately, using the quadratic formula.

    b. Use your expressions from part (a) to find, in simplified form, an expression for the sum of the solutions and an expression for the product of the solutions of a general quadratic equation.

9. $1 + \sqrt{2}, 1 - \sqrt{2}$

10. $\dfrac{3 + \sqrt{5}}{2}, \dfrac{3 - \sqrt{5}}{2}$

11. $\dfrac{-4 + 3\sqrt{2}}{2}, \dfrac{-4 - 3\sqrt{2}}{2}$

Use the result of part (b) of Exercise 8 to find _b_ and _c_ such that the equation $x^2 + bx + c = 0$ has the given solutions.

★ **12. a.** Suppose $2 + \sqrt{3}$ is a solution of a quadratic equation of the form $x^2 + bx + c = 0$, in which b and c are *integers*. Make a conjecture about the other solution of the equation, and find b and c based on your conjecture, using the results of part (b) of Exercise 8.

    b. Suppose $p + q\sqrt{r}$ is one solution of $x^2 + bx + c = 0$, where b and c are integers. What value must b have, in terms of p, q, and r, in order that $\sqrt{r}$ cancel out of the first two terms of the equation when $p + q\sqrt{r}$ is substituted for x?

    c. Use your answer to part (b), together with the result of part (b) of Exercise 8, to justify the assertion that $p - q\sqrt{r}$ is the other solution of the equation.

Challenge Problems, ALGEBRA 2: EXPLORATIONS AND APPLICATIONS

Challenge Set 27 ···

FOR USE WITH SECTION 5.6

Solve each equation. Check each solution.

1. $2^{x^2 + 2x} = 8$

2. $10^{x^2 - 3x} = \dfrac{1}{100}$

3. $2e^{2x} - 13e^x + 15 = 0$

4. $\log(x - 1) + \log(x + 2) = 1$

5. $\dfrac{3}{(x-2)^2} + \dfrac{11}{(x-2)} - 4 = 0$

6. $2x - 3 = \dfrac{20}{x}$

Solve each equation by factoring.

7. $5x(x - 8)(x + 2) = 0$

8. $-2(x + 5)(x - 3)(x + 7) = 0$

9. $2x^3 + 7x^2 - 4x = 0$

10. $x^2(x + 3) - x(x + 3) - 6(x + 3) = 0$

11. $x^2(x - 5) - 9(x - 5) = 0$

12. $x^4 - 34x^2 + 225 = 0$

13. $8x^6 + 7x^3 - 1 = 0$

14. $(x^2 - 4)^2 - 8(x^2 - 4) + 15 = 0$

Write each quadratic function as a product of factors, in terms of *a*, *b*, or both.

15. $y = ax^2 + (a + b)x + b$

16. $y = x^2 + (b - a)x - ab$

17. $y = abx^2 - (a + b)x + 1$

18. $y = ax^2 + (1 - ab)x - b$

★ **19. a.** Suppose a quadratic expression $x^2 + bx + c$, with b and c integers, can be factored. Explain how you know that the solutions of the equation $x^2 + bx + c = 0$ are integers.

   b. Suppose the solutions of a quadratic equation $ax^2 + bx + c = 0$, with a, b, and c integers and $a \neq 0$, are the rational numbers $\dfrac{p}{q}$ and $\dfrac{r}{s}$. What are the factors of $ax^2 + bx + c$?

   c. Based on your answer to part (b), make a conjecture about the relationship between the integer a and the integers q and s. Make a conjecture about the relationship between the integer c and the integers p and r.

20. Explain how you know that a quadratic expression of the form $ax^2 + bx$, with $a \neq 0$, can always be factored. Write the factors in terms of a and b.

Challenge Problems, ALGEBRA 2: EXPLORATIONS AND APPLICATIONS

31

Chapter 5 Challenge Set ·············

1. a. Suppose the parabola whose equation is $y = x^2$ is translated h units to the right and k units up. Write the equation of the new parabola in vertex form.

b. Find the x-coordinate of the intersection of the two parabolas, in terms of h and k.

c. Find values of h and k so that the two parabolas intersect on the line $x = 4$. (*Section 5.2*)

2. A *golden rectangle* has the property that a square (shaded in the diagram) can be removed from the rectangle and the small rectangle remaining is similar to the original rectangle.

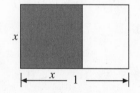

a. Suppose, as in the diagram, that the original rectangle has length 1. Set up a quadratic equation, in the variable x, based on the relationship of corresponding sides of similar polygons.

b. The positive solution of the quadratic equation you found in part (a) is called the *golden ratio*, often denoted by the Greek letter φ ("phi"). Find this solution to 3 decimal places.

c. Enter the number φ on a calculator. Press the reciprocal key. How do the numbers on the display change? Explain this result in terms of the equation that is satisfied by φ. (*Section 5.5*)

3. The twin towers of an office building are to be built on a plot of land that measures 250 ft by 160 ft. The floor plan of the towers will consist of two rectangles (shaded in the diagram), and a sidewalk of uniform width will surround the two rectangles, as shown.

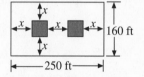

a. Let x = the width of the sidewalk. Write an equation that stipulates that the floor plan of the buildings will have a total area of 16,000 ft^2.

b. Solve for x under the condition in part (a). Is there more than one positive solution? If so, are both solutions physically possible? (*Section 5.6*)

4. *Extension* Try the following experiment using several different quadratic functions $f(x) = ax^2 + bx + c$.

First, evaluate each function at some integer values . . . –1, 0, 1, 2, 3, . . . and then find all the differences between these numbers: . . . $f(0) - f(-1)$, $f(1) - f(0)$, $f(2) - f(1)$, $f(3) - f(2)$, . . . Call these numbers the *first* differences of the function. Next find the differences between each pair of first differences. Call these numbers the *second* differences, and so on. Describe any patterns you find. Make a conjecture about the first and second differences of any quadratic function.

Challenge Set 28 ··

The table below lists hourly pay (in U. S. dollars) for production workers in selected countries for selected years. Use these data in Exercises 1–5.

Country	1975	1980	1985	1990
Japan	3.00	5.52	6.34	12.80
Mexico	1.44	2.21	1.59	1.64
Norway	6.77	11.59	10.37	21.47
Taiwan	0.40	1.00	1.50	3.95
U. S.	6.36	9.87	13.01	14.91

1. If the data for one country were to be displayed by year, would the year numbers represent categorical or numerical data?

2. What kind of graph should you use to display one country's data? Would it be possible to alter your graph so that data for all 5 countries could be displayed on one graph? If so, how could you do this?

3. Graph the data for each country in separate graphs. What trends do you notice within the data for one country? What comparisons can you make between the trends in different pairs of countries?

4. If you wanted to compare hourly pay in different countries for one particular year, would the data be categorical or numerical? What kind of graph would be appropriate for displaying such data?

5. Graph the data for all countries for the year 1990. What does your graph tell you that was not as apparent in the graphs of each country by year?

The table below gives the percents for the most popular intended majors among incoming college freshman in 1993. Use this table in Exercises 6 and 7.

Major	Health Related Majors	Business & Accounting	Education	Liberal Arts	Other
Percent of Freshmen	15.8%	13.1%	6.8%	12.9%	51.4%

6. Suppose that at a university the incoming freshman class of 1250 students declared the following majors: health related majors: 240; business and accounting: 140; education: 105; liberal arts: 180; other: 585. Use these figures to estimate the percent of incoming freshman in the United States who prefer each major. What type of graph would best display these data?

7. Compare the data for the university mentioned in Exercise 6 with the data in the table above. What conclusions might be drawn about trends since 1993?

Challenge Set 29 ···

In Exercises 1–5, the writer of the question has made an implication, which may or may not be true. (This is called a *loaded question*.) State this implication, give a reason why it might not be true, and rewrite the question without making the implication.

1. "Do you think our state legislators should be paid in accordance with the average salaries of similar full-time jobs?"

2. "Do you believe dogs should be allowed to run freely in the public parks as long as their owners keep them under control?"

3. "Would you be willing to pay an extra $1 on your tax return to finance cleaner political campaigns?"

4. "Do you support the unrestricted use of public transportation funds to finance worthy government projects?"

5. "Would you pay $25 a year for a subscription to the newsletter and join the ranks of those pictured above who have beaten the stock market?"

In Exercises 6–9, insert *one word* into each question (a) so that it is biased toward a "yes" answer, and (b) so that it is biased toward a "no" answer.

6. "Do you favor building more roads that would redirect through traffic around the downtown area?"

7. "Would you support a referendum to add the chipmunk to the national list of endangered species?"

8. "Do you think that the term of the county commissioner should be extended from four years to six years?"

9. "Would you favor the launching of another unmanned scientific probe to the outer planets of the solar system using tax dollars?"

In an election district, it is found that as-yet-unregistered voters would be more likely to vote for Democratic candidates than for Republican candidates.

10. Write a survey question that would ask residents of the district if they favor a voter registration drive in a way that would be more likely to extract a "yes" answer.

11. Ask the same question as in Exercise 10, again trying to extract a "yes" answer, but this time by including a technical term that residents will probably not understand.

Challenge Set 30 ···

FOR USE WITH SECTION 6.3

1. A manufacturer of liquid soap wants to find out whether her product will sell in a particular town. She wants to conduct a telephone survey, asking whoever answers the phone whether they would use a liquid soap. Describe a kind of sample that might be influenced by each factor.

   a. time of day   **b.** day of the week   **c.** the weather

2. Suppose an archaeological site is marked off with 100 small squares as shown at the right, and these squares are numbered from 1 to 100, starting at the upper left and proceeding across each row, from left to right, and proceeding downward row by row. A team of archaeologists wants to excavate a sample of the small squares. Tell what kind of sample would be produced by choosing the squares whose numbers are specified.

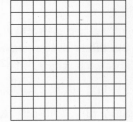

   a. numbers between 16 and 23   **b.** multiples of 4

   c. prime numbers   **d.** numbers less than 10

3. Which method of sampling in Exercise 2 might produce the least biased sample? Explain.

A calculator's random number generator "rand" produces a random number between 0 and 1. You can use this feature to produce a random integer between 1 and 10, for example, by entering the command int(10 * rand) + 1. In Exercises 4–7, tell how to use the calculator's random number generator to produce the specified kind of random number.

4. a random even integer between 1 and 20

5. a random odd integer between 1 and 100

6. a random 2-digit number whose first digit is 7

7. a random 2-digit number whose digits are the same

★ **8. a.** Describe a reasonably reliable method of using the square root key on your calculator to choose a sequence, of any desired length, of random integers between 1 and 9, inclusive. (*Hint*: Start with any number that is not a perfect square.)

   b. How can the method of part (a) break down?

   c. Modify the method of part (a) to produce integers between 1 and 99.

Challenge Set 31 ....

1. a. Of the three measures of central tendency—mean, median, and mode—which one can most easily be read from a histogram?

b. Suppose a relative frequency histogram has intervals consisting of single numbers, and suppose you multiply these numbers by their corresponding relative frequencies and add the results. Do you get one of the measures of central tendency mentioned in part (a)? If so, which one?

During two different time periods, the manager of a supermarket kept track of the times customers waited in the check-out lines before paying for their groceries. The numbers of minutes are given below.

Friday, 10:00-11:00 A.M.: 5, 6, 8, 5, 4, 3, 7, 10, 8, 6, 2, 3, 4, 7, 9, 11, 4, 3, 6, 5

Saturday, 4:00-5:00 P.M.: 4, 7, 9, 8, 8, 10, 12, 11, 10, 6, 9, 11, 10, 13, 12, 8, 9, 7, 5, 9

2. Make up a frequency table from each set of data, using intervals of 2 min.

3. Make a relative-frequency histogram from each set of data.

4. Predict the nature of the relative frequency histogram that would result from combining all the data into one set. Make a relative frequency histogram using the combined data. Was your prediction correct?

5. A consumer advocate found the following prices (in dollars) of the same automobile tire at different tire dealers: 45, 52, 50, 48, 55, 48, 49, 54, 51, 50.

a. Suppose two more dealers report their prices for the tire, and the median of all the prices is then $50. Can you find the prices reported by the two new dealers? If so, what are they? If not, what is your best estimate of what these prices are?

b. Is it possible that after two more dealers reported their prices, the median of all prices was $49.50? If so, show how this could happen. If not, explain why not.

6. A stem-and-leaf plot is a graphical representation of data in which all but the last digit of each data point (the stem) is listed to the left of a vertical line, and the last digits are listed to the right (the leaves) next to the corresponding first digit(s). For example, the data 25, 28, 32, 32, 34, 35, 40, 41, 44, would be displayed as

$$
\begin{array}{c|l}
2 & 5,\,8 \\
3 & 2,\,2,\,4,\,5 \\
4 & 0,\,1,\,4
\end{array}
$$

a. Construct a stem-and-leaf plot of the following data: 110, 114, 117, 118, 125, 128, 131, 132, 133, 133, 135, 142, 144, 145.

b. *Writing* Describe the similarities and differences between a stem-and-leaf plot and a histogram.

Challenge Set 32 ··

FOR USE WITH SECTION 6.5

1. The table below lists the frequencies of the percentage rates of sales tax in those states that have a sales tax.

Percent Rate	Number of States
3%	3
4%	9
4.5%	2
5%	15
6%	13
6.5%	2

a. Find the standard deviation of the data.

b. Suppose in a set of data f_1 is the frequency of x_1, f_2 is the frequency of x_2, and so on. Modify the formula for the standard deviation of a set of data so that it is given in terms of the data points and their respective frequencies.

2. Suppose the mean of a set of data is $\bar{x}$ and its standard deviation is σ.

a. Suppose that every data point is then multiplied by a number k. Find the mean and the standard deviation of the new data in terms of $\bar{x}$, σ, and k.

b. Suppose that a number k is added to every data point in the original data set. Find the mean and the standard deviation of the new data in terms of $\bar{x}$, σ, and k.

★ **3. a.** The number under the square root sign in the formula for σ is called the *variance* of a set of data and is often denoted σ^2. Use the standard deviation formula to show that the variance can be written

$$\sigma^2 = \frac{x_1^2 + x_2^2 + \ldots + x_n^2}{n} - 2\bar{x}\left(\frac{x_1 + x_2 + \ldots + x_n}{n}\right) + (\bar{x})^2.$$

b. Simplify the formula you found in part (a) by writing $\overline{x^2}$ for the mean of the numbers $x_1^2, x_2^2, \ldots, x_n^2$. Use this simplification to write another formula for σ.

4. Find the value of x so that the set of data, 9, 5, 3, 2, x has standard deviation $\sqrt{6}$. (*Hint*: Use the formula for s that you found in part (b) of Exercise 3.)

Challenge Set 33 ···

FOR USE WITH SECTION 6.6

The number $2\sqrt{\dfrac{\hat{p}(1-\hat{p})}{n}}$, where $\hat{p}$ is the proportion of a sample having a

given characteristic and n is the size of the sample, gives a better estimate of
the margin of error for values of $\hat{p}$ that are not close to 0.5 than does the

number $\dfrac{1}{\sqrt{n}}$. In Exercises 1 and 2, use this estimate to find each margin of

error. Also find the margin of error using $\dfrac{1}{\sqrt{n}}$.

1. A survey that found that 73 people out of a sample of 100 favored a gasoline tax decrease.

2. A telephone poll found that 64 people out of 350 favored using public funds to support replacing a downtown sidewalk.

3. Suppose another survey among a sample of 550 people also asked whether respondents favored a gasoline tax decrease, and suppose this survey had a margin of error of 4%. How many people in the sample favored the decrease?

4. A polling organization conducted a poll of a sample of the population of a city and found, using the formula above, that between 58% and 70% of the population of the city thought that police protection in the city was adequate. How many voters participated in the survey?

5. A market researcher for a cereal manufacturer believes that about 50% of the population would prefer her company's product to a competitor's product. She wants to conduct a taste test to find if this is true . The researcher wants the test to provide an estimate that is accurate to within ±5%. How large a sample will be required?

6. By telephoning sample lists of voters consisting of 100 voters each, a pollster found the following numbers of voters in the town favoring Ramirez, one of the candidates for mayor: 54, 48, 44, 56, and 52.

    a. Let $\hat{p}$ = the *mean* of the proportions of the samples that favor Ramirez. What mean margin of error would the formula above predict for each of the samples?

    b. Using the formula above and the value of $\hat{p}$ from part (a), what margin of error would be expected if a sample of 400 voters was polled? Answer the same question for a sample of 1600. Describe any pattern you notice.

★ 7. *Writing* Describe the values produced by the formula above for the margin of error when $\hat{p}$ is very close to 1 or very close to 0. Regarding this formula as a function of $\hat{p}$, $f(\hat{p})$, what is the relationship between $f(\hat{p})$ and $f(1-\hat{p})$? Prove this relationship.

Chapter 6 Challenge Set ·······························

1. A computer is to be programmed to select 20 students at random from among the 420 students in the senior class. The computer contains a numbered list of the students, and the following method is to be used: The computer will choose a random positive integer n, then it will count through the students, starting over, if necessary, when it reaches the end of the list, and select the student whose number it lands on. It will then count n students from that point in the list, again selecting the student it lands on, and it will continue in this way until all 20 students are selected. (*Section 6.2*)

   a. Give three possible values of n for which this procedure would break down. Which values of n should probably be rejected if the computer chooses them?

   b. Find values of n that would cause the procedure to produce each of the following: a cluster sample, a systematic sample, a random sample.

2. The owner of a Mexican restaurant conducted a survey and found that 75 of those people surveyed preferred Mexican food. Using the formula

$$\text{margin of error} = 2\sqrt{\frac{\hat{p}(1-\hat{p})}{n}},$$

the restaurant owner calculated that there was a 5% margin of error in the survey. Find the number of people in the survey. (*Hint*: $\frac{75}{n}$ is the sample proportion $\hat{p}$. Find a polynomial equation involving n and solve it by graphing.) (*Section 6.6*)

3. *Extension* The following list gives the earned-run averages of the winners of the American League and National League Cy Young Awards in major league baseball between 1967 and 1993, excluding 1981. (There was a tie in the American League in 1969.)

AL: 3.16, 1.96, 2.38, 2.80, 3.03, 1.82, 1.92, 2.40, 2.49, 2.09, 2.51, 2.17, 1.74, 3.08, 3.23, 3.34, 3.66, 1.92, 2.87, 2.48, 2.97, 2.64, 2.16, 3.06, 2.62, 1.91, 3.37

NL: 2.85, 1.12, 2.21, 3.12, 2.77, 1.97, 2.08, 2.42, 2.38, 2.74, 2.64, 2.72, 2.23, 2.34, 3.10, 2.37, 3.64, 1.53, 2.22, 2.83, 2.26, 1.85, 2.76, 2.55, 2.18, 2.36

For each data set, find the mean $\bar{x}$ and the standard deviation σ, and fill in the correct *percentages* in the following table.

	$x \le \bar{x} - 2\sigma$	$x \le \bar{x} - \sigma$	$x < \bar{x}$	$x \le \bar{x} + \sigma$	$x < \bar{x} + 2\sigma$	$x < \bar{x} + 3\sigma$
American						
National						

From your table, describe how to find the percentage of data that falls between two multiples of the standard deviation. Find this percentage for $\bar{x} - \sigma < x \le \bar{x} + \sigma$ and for $\bar{x} - 2\sigma < x \le \bar{x} + 2\sigma$. Compare your results with the theoretical results given on page 269 of the textbook.

Challenge Set 34 ···

Solve each system of equations. (*Hint*: Substitute *u* and *v* for expressions involving *x* and *y*, respectively, and solve the resulting *u-v* system.)

1. $\left(\dfrac{1}{x}\right) - 3\left(\dfrac{1}{y}\right) = 9$

$2\left(\dfrac{1}{x}\right) + \left(\dfrac{1}{y}\right) = 4$

2. $4x^2 + 5y^2 = 30$

$7x^2 + y^2 = 37$

3. $2\,|x - 1| - |y + 3| = -1$

$5\,|x - 1| - 2\,|y + 3| = 1$

Solve each system algebraically.

EXAMPLE: $y = x^2 - 3$
$y = 2x - 3$

SOLUTION: Substitute the expression for *y* from the second equation into the first:

$2x - 3 = x^2 - 3 \;\Rightarrow\; 0 = x^2 - 2x \;\Rightarrow\; 0 = x(x - 2) \;\Rightarrow\; x = 0 \text{ or } 2.$
Substituting $x = 0$ in either original equation gives $y = -3$; substituting $x = 2$ gives $y = 1$. Solutions are $(0, -3)$ and $(2, 1)$.

4. $y = 2x^2 - 3x + 1$

$y = -x + 5$

5. $y = \dfrac{1}{2}x^2 + x - 5$

$y = 2x - 1$

6. $y = -x^2 + 5x + 7$

$y = 3x + 4$

7. Solve each system by substituting an expression for *z* obtained from the first equation into each of the other two equations.

a. $x + y + z = 4$
$2x - y + z = 3$
$-x + 3y - 2z = 5$

b. $3x + 2y + z = 8$
$x - y + 3z = 5$
$2x + y + 2z = 9$

8. A football team is attempting a field-goal from the 15-yard line (75 ft from the goal post). The path of the ball is given by the parametric equations $x = 60t$, $y = 28t - 16t^2$, where *t* is the time in seconds after the ball is kicked, and the origin of the coordinate system is at the kicking tee, with *x* and *y* measured in feet.

a. Find a relationship between *x* and *y* that is true for every point on the path of the ball.

b. Will the ball go over the goal-post crossbar, which is 10 ft high?

★ **9.** Geometrically, a line will be tangent to a parabola if and only if the solution of the corresponding system of equations contains only one point (x, y). Use this fact to show algebraically that the line $y - a^2 = 2a(x - a)$ is tangent to the parabola $y = x^2$ at the point (a, a^2).

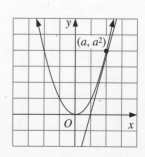

Challenge Set 35 ···

FOR USE WITH SECTION 7.2

Solve each system.

1. $3\left(\dfrac{1}{x-4}\right) + 2\left(\dfrac{1}{y+1}\right) = 11$

$5\left(\dfrac{1}{x-4}\right) - 4\left(\dfrac{1}{y+1}\right) = 33$

2. $4e^x - 5e^y = 1$

$6e^x - 2e^y = 18$

3. $\log x^3 - \log y^2 = 8$

$\log x^4 + \log y^6 = 2$

4. $2^{5x+2y} = 16$

$2^{4x+y} = \dfrac{1}{2}$

5. When one billiard ball A collides head-on with another billiard ball B that is at rest, two quantities are conserved: momentum (mass $\times$ velocity) and kinetic energy ($\dfrac{1}{2}$ mass $\times$ velocity^2). This means that the sum total of each of these quantities for the whole system (both billiard balls) is the same before and after the collision. Suppose both balls have mass m, and the velocity of ball A *before* the collision is v_0. Let v and w be the unknown velocities of balls A and B, respectively, *after* the collision.

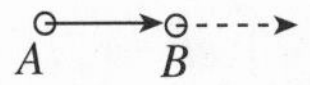

   a. Write two equations that assert that each of the two quantities above is conserved.

   b. Eliminate v_0 from the system in part (a) to get a single equation involving v and w. Find two possible solutions for v and w using this equation and the momentum equation you found in part (a).

   c. Explain why one of the solutions in part (b) is not possible *after* the collision. (*Hint:* Refer to the diagram above.) Describe what happens physically after the balls collide.

★ **6.** The determinant of a 2×2 matrix $A = \begin{bmatrix} a & b \\ c & d \end{bmatrix}$ is $ad - bc$, denoted det A.

   You can use determinants to solve a system of linear equations

$$ax + by = p$$
$$cx + dy = q$$

   by the following method, called *Cramer's Rule.* Let $B = \begin{bmatrix} p & b \\ q & d \end{bmatrix}$ and

   $C = \begin{bmatrix} a & p \\ c & q \end{bmatrix}$. Then the solution of the system above is given by

$$x = \frac{\det B}{\det A} \qquad y = \frac{\det C}{\det A}.$$

   Prove that this method works by solving for x and y using addition or subtraction.

Challenge Problems, ALGEBRA 2: EXPLORATIONS AND APPLICATIONS

Challenge Set 36 ·······································

1. The *determinant* of a 3×3 matrix $A = \begin{bmatrix} a & b & c \\ d & e & f \\ g & h & k \end{bmatrix}$ is the number $aek + bfg +$ $cdh - gec - hfa - kdb$, denoted det A.

a. Show that if A is a *diagonal matrix*, (that is, A has the form $\begin{bmatrix} a & 0 & 0 \\ 0 & e & 0 \\ 0 & 0 & k \end{bmatrix}$), then det A is the product of the diagonal entries.

b. Show that if A is a diagonal matrix with det $A \neq 0$ and $B =$

$\dfrac{1}{\det A} \begin{bmatrix} ek & 0 & 0 \\ 0 & ak & 0 \\ 0 & 0 & ae \end{bmatrix}$, then $AB = I$, the 2×2 identity matrix. What does this

tell you about B?

2. a. A matrix $A = \begin{bmatrix} a & b & c \\ 0 & e & f \\ 0 & 0 & k \end{bmatrix}$ is called a *triangular matrix*. Use the formula in

Exercise 1 to show that the determinant of a triangular matrix is the product of the diagonal entries.

b. Let A be the matrix in part (a) (det $A \neq 0$) and let $B = \dfrac{1}{\det A} \begin{bmatrix} ek & -bk & bf - ce \\ 0 & ak & -af \\ 0 & 0 & ae \end{bmatrix}$. Show that $AB = I$, the 3×3 identity matrix.

3. Use the formula given in Exercise 2 to find the inverse of each matrix.

a. $\begin{bmatrix} -1 & -4 & 3 \\ 0 & 1 & -2 \\ 0 & 0 & 1 \end{bmatrix}$ **b.** $\begin{bmatrix} -1 & 3 & 4 \\ 0 & 2 & -2 \\ 0 & 0 & 1 \end{bmatrix}$

★ **4.** Let $A = \begin{bmatrix} a & b & c & p \\ d & e & f & q \\ g & h & k & r \end{bmatrix}$ be associated with the system $\begin{cases} ax + by + cz = p \\ dx + ey + fz = q \\ gx + hy + kz = r \end{cases}$. By adding

a multiple of one row to another row, you get an *equivalent* matrix. For example,

if you add -3 times the 1st row of $\begin{bmatrix} 1 & 2 & 3 & 20 \\ 0 & 4 & -1 & 8 \\ 3 & 6 & 5 & 44 \end{bmatrix}$ to the 3rd row, you get $\begin{bmatrix} 1 & 2 & 3 & 20 \\ 0 & 4 & -1 & 8 \\ 0 & 0 & -4 & -16 \end{bmatrix}$.

If you get such an equivalent matrix, the solution can be read easily. For example, the system is now $x + 2y + 3z = 20$, $4y - z = 8$, $-4z = -16$, and it is easy to find z, then y, then x. Solve each system.

a. $\begin{bmatrix} 2 & -1 & 4 & 2 \\ 4 & 3 & 1 & 17 \\ 2 & -1 & -3 & 5 \end{bmatrix}$ **b.** $\begin{bmatrix} 3 & -1 & 0 & 1 \\ -9 & 6 & 2 & 4 \\ 6 & -2 & 1 & -2 \end{bmatrix}$

Challenge Set 37 ·· ···················

Graph each inequality.

1. $y \geq |x - 3|$

2. $y < 8 - x^2$

3. $y < x^2 - 6x + 9$

4. $y \leq 2 \log x$

5. $y \geq 2^x - 1$

6. $y > \frac{1}{2}|x + 3| + 1$

Find an inequality that fits each description.

7. The graph of the inequality contains the points (3, 4) and (–1, 10) on its boundary and the point (1, 8) in its interior.

8. The boundary of the graph of the inequality contains (–2, 3) and (4, 5), but these points are not in the graph. The graph contains the origin.

9. The graph of the inequality consists of all points that are on or below the line passing through (1, –2) and having slope $\frac{2}{5}$.

10. The graph consists of all points above the line with positive slope that passes through the point (3, –2) and makes a 45° angle with the x-axis.

11. a. Describe or draw accurately a region R of the coordinate plane that has the following property: If l is any line through the origin, the set of points common to the line l and the region R is a ray with endpoint at the origin.

   b. Describe the region R in part (a) using one or more inequalities.

Graph each inequality.

12. $|x| + |y| \leq 9$

13. $|x - y| < 3$

14. $|x + y| \geq 2$

★ **15. a.** Graph the inequality $y > \frac{1}{2}x + 6$.

   b. On the same axes graph the parametrically defined curve $x = 2t$, $y = 12 - t^2$. Find the values of t corresponding to those points on the curve that lie in the region you graphed in part (a).

★ **16.** Cardiss and Inez live in cities that are 60 mi apart along a straight road. Each morning at 8:00 A.M. the two women set out by car along the road toward the other city, both traveling at 45 mi/h. Let t = the time (in hours) after 8:00 A.M. Graph those points (t, y) such that y is greater than the distance between the two cars at time t for the time they are on the road.

Challenge Set 38 ···

Graph each system of inequalities.

1. $y < 5 - x^2$
$y > x^2 - 3$

2. $y \geq x$
$y \geq -x$
$y \leq x + 4$
$y \leq 4 - x$

3. $y > x^2 - 2x$
$y < 3x^2 - 6x$

4. $y \leq 2^x + 5$
$y \leq 2^{-x} + 5$

5. $y > |x - 4| - 2$
$y < -|x - 3| + 3$

6. $|x| < 4$
$x - 2 < y < x + 2$

7. $y < x + 2$
$y < 2 - x$
$y > x^2 - 4$

8. $1 \leq |y - 2| \leq 3$
$1 \leq |x - 1| \leq 3$

9. $x \geq 0$
$y \geq 0$
$y \leq -\dfrac{1}{2}x + 5$
$y \leq -2x + 11$

Write a system of inequalities defining each shaded region.

10.
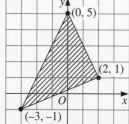

11.
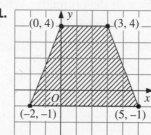

12.
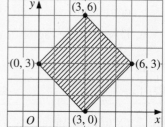

13. An isosceles triangle has its vertex at the point (2, 5) and has a base of length 4. The altitude of the triangle drawn from the vertex also has length 4 and is contained in a horizontal line. Find a system of inequalities that defines the interior (not including the sides) of this triangle.

★ **14.** Girija and Han-Ling decide to meet at a restaurant for dinner if they finish work in time. Each woman will arrive at the restaurant at some time between 5:00 P.M. and 6:00 P.M., and they have agreed that whoever arrives first will wait 10 minutes or until 6:00 P.M. (whichever comes first) for the other woman to arrive, and will leave the restaurant if she does not.

    a. In a coordinate plane, let each point (x, y) represent a pair of possible arrival times (in minutes after 5:00 P.M.) of the two women at the restaurant, the x-coordinate representing Girija's arrival time and the y-coordinate representing Han-Ling's arrival time. Shade in the region of the coordinate plane that represents the pairs of arrival times for which the two will meet each other.

    b. Write a system of inequalities that defines the region of the coordinate plane that you shaded in part (a).

Challenge Set 39 ··

FOR USE WITH SECTION 7.6

1. A factory that makes in-line skates has two models: the Cougar and the Piranha, both of which require the work of 3 machines. A cuttihg machine needs to spend 10 min on each pair of the Cougar model skates and 20 min on each pair of the Piranha model; a stitching machine needs to spend 20 min on each pair of both models; and a finishing machine needs to spend 30 min on each pair of the Cougar model and 7.5 min on each pair of the Piranha model. There is enough skilled labor at the factory to run the cutting machine for 8 h per day, and each of the other two machines for 9 h per day.

    a. Let x = the number of Cougar model skates the factory produces per day, and let y = the number of Piranha model skates produced per day. Write a system of inequalities that expresses the constraints on x and y. (*Hint*: Convert each time given in minutes to a fraction of an hour.)

    b. Graph the system of inequalities you wrote in part (a).

    c. Suppose the company makes a profit of $30 on each pair of Cougar model skates and $20 on each pair of the Piranha model. Find the values of x and y that satisfy the constraints in part (a) and produce the maximum profit for the company.

2. *Writing* Use the diagram at the right to explain why the function $x + y = n$ cannot be maximized at an interior point of a convex polygon. (*Hint*: Suppose $x + y = n$ is maximized at an interior point $P(a, b)$ in the first quadrant, with $a + b = k$. There must be a point of the form $(a + c, b)$, for some $c > 0$, that is also an interior point.)

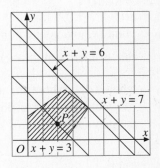

★ **3. a.** Suppose a feasible region for a system of inequalities is partly bounded by a parabola, as shown, and suppose the maximum value of $x + y$ in the region occurs at a point $P(x, y)$ on the parabola, where $x + y = k$. Explain why the graph of the line $x + y = k$ must be *tangent* to the parabola at P.
(*Hint*: Suppose this line were not tangent.)

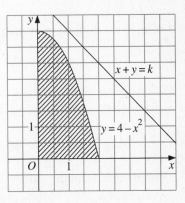

    b. Find a system of inequalities that defines the feasible region shown.

    c. A tangent to the parabola $y = 4 - x^2$ at the point (a, b) has slope $-2a$. Use this fact to find the point at which the maximum value of $x + y$ in the feasible region occurs. What is this maximum value?

Chapter 7 Challenge Set ·······················

1. Juanita put her canoe in a river at the boat landing and rowed 1 mi upstream, where her hat was blown overboard. Juanita, however, was watching a bird and did not notice the loss of the hat. One hour later she did notice, and immediately turned the canoe around and rowed back downstream, catching up with her hat right at the boat landing. (*Section 7.1*)

a. Let x = Juanita's paddling speed relative to the water, and let y = the speed of the current. Write an expression, involving x, y, or both for the *distance* Juanita rowed during the hour between the time she lost her hat and the time she discovered it was missing.

b. Using the distance you found in part (a), write an expression in x and y for the time Juanita rowed after the moment when the hat was blown overboard. (*Hint*: This time is the time she rowed upstream + the time she rowed back downstream. Time = distance ÷ rate.)

c. Write an expression for the time the hat was in the water. By equating this expression with the expression you found in part (b), find y. Can you also find x?

2. A matrix $\begin{bmatrix} a\,b \\ c\,d \end{bmatrix}$ can be viewed as a *linear transformation* that assigns to a point (x, y) a point (x', y') in a different coordinate plane, according to the rule of matrix multiplication: $\begin{bmatrix} a\,b \\ c\,d \end{bmatrix}\begin{bmatrix} x \\ y \end{bmatrix} = \begin{bmatrix} x' \\ y' \end{bmatrix}$. For each matrix, draw a labeled diagram of the image of the square with vertices (0, 0), (1, 0), (1, 1), (0,1) under the corresponding linear transformation, and describe in words the geometric effect of the transformation. (*Section 7.3*)

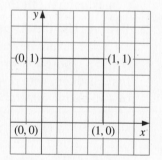

a. $\begin{bmatrix} 0\,1 \\ 1\,0 \end{bmatrix}$ **b.** $\begin{bmatrix} 1\,1 \\ 0\,1 \end{bmatrix}$ **c.** $\begin{bmatrix} -1\,0 \\ 0\,1 \end{bmatrix}$ **d.** $\begin{bmatrix} 0\,-1 \\ 1\,0 \end{bmatrix}$

3. Graph the following system of inequalities. (*Section 7.5*)

$$|x + y| \le 3$$
$$|x - y| \le 3$$

4. *Extension* Suppose the shaded region at the right is the feasible region for a system of constraints, and a function of the form $ax + by$ is to be maximized on this region. Find the point(s) at which each of the following functions has its maximum on the feasible region: $3x + y$, $2x + y$, $x + y$, $3x + 4y$, and $x + 4y$. From your results, find a way to predict, from the values of a and b, without actually substituting in the values of x and y at the corner points, where the maximum of $ax + by$ will occur. Test your conjecture with some functions.

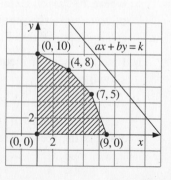

Challenge Set 40 ···

Find the inverse function of each given function.

1. $f(x) = \sqrt{x} - 3$

2. $f(x) = x^2 + 1,\ x \le 0$

3. $f(x) = -\sqrt{x+5} + 3$

4. $f(x) = (x-4)^2 + 1,\ x \ge 4$

5. $f(x) = \sqrt{2x-7} - 3$

6. $f(x) = 2(x+3)^2 - 4,\ x \ge -3$

7. Explain how you know that $\sqrt{x^2} = |x|$.

8. Use the fact stated in Exercise 7 to simplify each expression.

   a. $\sqrt{50x^2}$ **b.** $\sqrt{48y^3}$ **c.** $\sqrt{500pq^2}$ **d.** $\sqrt{490a^4}$

Solve each inequality.

   EXAMPLE: $\sqrt{x+2} \le 5$

   SOLUTION: Since both sides are positive, the inequality is preserved when
you square both sides: $\sqrt{x+2} \le 5 \ \Rightarrow\ x + 2 \le 25 \ \Rightarrow\ x \le 23$
The solution is $-2 \le x \le 23$.

9. $\sqrt{x-3} - 4 \ge 0$ **10.** $5 - \sqrt{x+4} \ge 2$ **11.** $2\sqrt{x-4} - 5 \le 0$

Graph each function. State the domain and range.

12. $y = \sqrt{|x-2|}$ **13.** $y = x|x| - 3$ **14.** $y = -\sqrt{2x+3} + 5$

15. a. Find two numbers a and b such that $\sqrt{a} + \sqrt{b} = \sqrt{a+b}$.

   b. Make a conjecture about what pairs (a, b) satisfy the equation in part (a).
Prove your conjecture by simplifying that equation. (*Hint*: Square both sides.)

16. Find a and b if the points $(6, 4)$ and $(17, 7)$ are on the graph of the function
$y = \sqrt{ax + b}$.

★ **17. a.** Find the domain and range of the function $f(x) = \sqrt{ax + b} + c$, in terms of
a, b, and c.

   b. Find f^{-1} in terms of a, b, and c.

   c. Find the vertex of the part of a parabola that is the graph of f^{-1}. Use your
answer to find the vertex of the "sideways" parabola that is the graph of f.

Challenge Set 41 ···

Simplify each expression. Assume all variables represent positive real numbers.

1. $\sqrt[3]{24x^3}$

2. $\sqrt[5]{96a^9}$

3. $\sqrt[3]{4y} \cdot \sqrt[3]{54y^5}$

4. $\dfrac{\sqrt[4]{3u^6v^2}}{\sqrt[4]{48u^2v^6}}$

5. $3\sqrt[6]{20x^4} \cdot \sqrt[6]{16x^8}$

6. $\dfrac{\sqrt[4]{5p^3q^8}}{\sqrt[4]{405p^7}}$

7. a. Simplify $\sqrt{\sqrt[3]{64}}$ by using fractional exponents.

★ **b.** Use fractional exponents to show that $\sqrt[m]{\sqrt[n]{x}} = \sqrt[mn]{x}$, for any positive number x and any positive integers m and n.

Graph each function.

8. $y = \sqrt[3]{|x|^2}$

9. $y = \sqrt{|x|^3}$

10. $y = \sqrt[3]{|x|^4}$

11. $y = \sqrt[4]{|x|}$

12. The shapes of the graphs in Exercises 8–11 fall naturally into two distinct groups geometrically. Which graphs belong to each group? If you were given another function of the same type, how could you tell which group its graph would fall in without actually graphing the function?

13. a. Explain how you know that, for any positive number x and any two positive integers p and q, $x^{pq} = x^{qp}$.

★ **b.** Prove that if m and n are positive integers and $x \geq 0$, then $\sqrt[n]{x^m} = \left(\sqrt[n]{x}\right)^m$.

Use the following method: Let $y = \sqrt[n]{x^m}$ and let $z = \left(\sqrt[n]{x}\right)^m$. Calculate y^n and z^n separately, and use the result of part (a).

14. a. On one coordinate system, sketch the graphs of $y = \sqrt{x}$, $y = \sqrt[4]{x}$, and $y = \sqrt[8]{x}$ for $0 \leq x \leq 1$. (*Hint*: Use large units.)

b. On the basis of the graphs you drew in part (a), describe what happens to the graph of $y = \sqrt[n]{x}$ for $0 \leq x \leq 1$ as n gets larger and larger. If a is a fixed value between 0 and 1, make a conjecture about what happens to the value of $\sqrt[n]{a}$ as n gets large.

Challenge Set 42 ·············· ········· ······

Solve each equation. Check to eliminate extraneous solutions.

EXAMPLE: $\sqrt{x+4} - \sqrt{x-1} = 1$

SOLUTION: Before squaring, get the radicals on opposite sides of the equation: $\sqrt{x+4} = \sqrt{x-1} + 1$

Square both sides: $x + 4 = \left(\sqrt{x-1}\right)^2 + 2\sqrt{x-1} + 1 \Rightarrow$

$x + 4 = x - 1 + 2\sqrt{x-1} + 1 \Rightarrow 4 = 2\sqrt{x-1}$

Square both sides again: $16 = 4(x-1) \Rightarrow 4x = 20 \Rightarrow x = 5$

1. $\sqrt{x+9} = \sqrt{x} + 1$

2. $3 - \sqrt{y-6} = \sqrt{y}$

3. $\sqrt{a+10} = 5 - \sqrt{a}$

4. $\sqrt{k} - \sqrt{k-8} = 2$

5. $\sqrt{r+10} + \sqrt{3-r} = 5$

6. $\sqrt{p+5} + \sqrt{p-2} = 7$

7. $3\sqrt{c-1} - \sqrt{c-8} = 9$

8. $5\sqrt{n+4} - \sqrt{n-1} = 11$

9. $\sqrt{2x+1} - \sqrt{x-3} = 2$

10. $\sqrt{b+3} + \sqrt{3b+10} = 3$

11. $\sqrt{u-2} - \sqrt{2u+3} + 2 = 0$

12. $\sqrt{2z+8} - \sqrt{z+2} = 2$

13. $\sqrt{3x+1} + \sqrt{x-1} = 6$

14. $\sqrt{2w+3} - \sqrt{w+1} = 1$

★ 15. $\dfrac{2d-3}{\sqrt{3d-2}} = \sqrt{d}$

★ 16. $\dfrac{3-2a}{\sqrt{2a-3}} - \sqrt{2a} + 2 = 0$

★ 17. $\sqrt[3]{x^3+7} - 1 = x$

★ 18. $\sqrt[3]{y^3-3y-1} = y - 1$

19. Odetta has created a fitness program for herself that includes swimming ashore from a boat mooring (P in the diagram) 0.6 mi from the shore. She then jogs to her house, which is 2 mi from the point (Q in the diagram) on the shore nearest to the mooring. Odetta swims at 2 mi/h and jogs at 6 mi/h. Her route is the heavy line in the diagram.

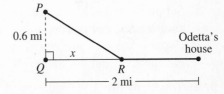

a. Let x = the distance from Q to the point R where she comes ashore. Write expressions, in terms of x, for the time she spends swimming and the time she spends jogging.

b. One day Odetta's combined time for swimming and jogging was 42 min. Find the distance x.

Challenge Set 43 ···

Solve each quadratic equation using the quadratic formula.

1. $x^2 - 6ix - 9 = 0$ **2.** $ix^2 - 3x - 2i = 0$ **3.** $3x^2 + 5ix + 12 = 0$

4. Show that $\dfrac{p + qi}{r + si}$ can be written in $a + bi$ form, for any real numbers p, q, r,

     and s, provided $r + si \neq 0$. Give the values of a and b in terms of r and s.

5. a. Suppose that for some real numbers a, b, c, and d, $(a + bi)^2$ equals $c + di$.
        Show that $(-a - bi)^2$ also equals $c + di$.

    b. Show that $4 - 3i$ is a square root of $7 - 24i$.

    c. Based on the result of part (a), what is another square root of $7 - 24i$?

6. Find i^3, i^4, i^5, and i^6. Make a conjecture about the nth power of i where n is an
    integer that gives a remainder of r when divided by 4 where $r = 0, 1, 2,$ or 3.

7. If $z = a + bi$, let $\bar{z}$ denote the conjugate $a - bi$ of z. Also, let $w = c + di$.
    Prove each of the following assertions.

    a. $\bar{z} + \bar{w} = \overline{z + w}$

    b. $\bar{z} \cdot \bar{w} = \overline{z \cdot w}$

    c. If k is a real number, then $k \cdot \bar{z} = \overline{kz}$. Why is this the same as $\bar{k} \cdot \bar{z} = \overline{kz}$?

8. a. Use the result of Exercise 7 to justify the following statement: If $P(x)$ is a
        polynomial <u>function</u> *with real coefficients*, $P(x) = a_n x^n + \ldots + a_1 x + a_0$,
        then $P(\bar{z}) = \overline{P(z)}$.

    b. Suppose $z = a + bi$ is a solution of a polynomial function P with real
        coefficients, that is, suppose $P(z) = 0$. Explain how you know that $\bar{z}$ is
        also a solution of this equation. (*Hint*: Take the conjugate of both sides of
        the equation and use the result of part (a).)

9. a. You know that 1 is a solution of the equation $x^3 - 1 = 0$. By direct

        calculation, show that $\frac{1}{2}(-1 + i\sqrt{3})$ is also a solution of this equation.

    b. Based on the result of part (b) of Exercise 8, what is the third solution of
        the equation $x^3 - 1 = 0$? By direct calculation, show that your answer is
        actually a solution.

Challenge Set 44 ....

1. Plot each given complex number z and, in the same complex plane, plot the complex number $-iz$. (*Hint*: Use equal horizontal and vertical units.)

a. $z = 3 + 4i$ **b.** $z = -2 + i$ **c.** $z = 5 - 3i$ **d.** $z = -1 - 4i$

2. Join each pair of points z and $-iz$ from Exercise 1 to the origin with line segments. What do you notice? Make a conjecture about multiplying any complex number by $-i$.

3. For each pair of numbers, accurately plot the numbers and their product in the same complex plane. Connect each point to the origin by a line segment and use a protractor to measure the angle each segment makes with the positive real axis. What relationship do you notice among the three angle measurements in each case?

a. $2 + i$, $3 + 2i$ **b.** $1 + i$, $3 + 4i$ **c.** $1 + 2i$, $1 + 3i$

4. Measure the magnitudes of the two complex numbers and their product that you plotted in parts (a) and (b) of Exercise 3. What relationship do you notice among the three magnitudes in each case?

5. Make a conjecture about how the angle and the magnitude of a *quotient* of two complex numbers relates to the angles and magnitudes of the two numbers.

6. Let $z = 1 + i$. Plot the numbers z, z^2, z^3, z^4, and z^5 in one complex plane and connect each plotted point to the origin with a line segment.

  a. Describe any pattern you notice among the angles these segments make with the positive x-axis. (*Hint*: Think of angles being measured in degrees counterclockwise from the positive real axis, so that an angle may have a measure greater than 180°.)

  b. Describe any pattern you notice among the magnitudes of these segments.

  c. Plot the points $z^0 = 1$ and $z^{-1} = \dfrac{1}{z}$. (*Hint*: First convert $\dfrac{1}{z}$ to $a + bi$ form.) Do these points continue the pattern you noticed in parts (a) and (b)?

  d. Predict the magnitude and angle of z^6. Plot this point. Then calculate its value to check whether you were correct.

7. *Writing* The points you plotted in Exercise 6 all lie on a spiral, called an *equiangular* spiral. Sketch this curve, and draw some additional lines radiating from the origin. What geometric property do you think accounts for the name of this spiral?

Challenge Set 45 ···

1. a. Show that there cannot be two distinct identity elements in a group; that is, suppose that each of two elements I and J has the property that when it is multiplied by an element A, the product is A. Then show that $I = J$. (*Hint*: Consider the product IJ.)

b. Suppose A is an element of a system with an associative multiplication operation defined. Suppose that B is an element of the system such that $BA = I$, the identity element, and C is an element (possibly different from B) such that $AC = I$. Show that $B = C$. (*Hint*: Consider the product $(BA)C$. Use associativity.)

2. The determinant of a 2 × 2 matrix $A = \begin{bmatrix} a & b \\ c & d \end{bmatrix}$ is $ad - bc$, denoted det A.

Consider the set S of all 2 × 2 matrices A such that det $A = 1$ or det $A = -1$.

a. Is the set S closed under the operation of matrix multiplication? (*Hint*: det $AB = ($det $A)($det $B).$)

b. What is the identity element under this multiplication?

c. Is this multiplication commutative? Explain why, or give a counterexample to show that it is not.

3. Let T = the set of all 2 × 2 matrices of the form $\begin{bmatrix} a & -b \\ b & a \end{bmatrix}$. Use this matrix and

another of the same form $\begin{bmatrix} c & -d \\ d & c \end{bmatrix}$ to answer each of the following questions.

a. Is the set T closed under the operation of matrix multiplication? Prove your answer.

b. Is multiplication of elements of T commutative? Prove your answer.

c. Does every element of T have an inverse with respect to matrix multiplication?
(*Hint*: The inverse of a 2 × 2 matrix $A = \begin{bmatrix} a & b \\ c & d \end{bmatrix}$ is $\frac{1}{\det A} \begin{bmatrix} d & -b \\ -c & a \end{bmatrix}$.)

d. *Writing* Let the first matrix in part (a) correspond to the complex number $a + bi$. What can you say about the correspondence between a product of two such matrices and the product of the two corresponding complex numbers? Compare the inverse of a matrix with the reciprocal of the complex number corresponding to that matrix.

★ **4.** Let Z_n be the set of integers 1, 2, 3, . . . , n with multiplication defined as on a clock having just these numbers. (For example, in Z_5, $3 \cdot 4 = 2$, since 2 is where you will end up if you start at 1 and count up to 12.) Does Z_n always constitute a group, no matter what n is? If so, explain why. If not, for which values of n is it a group?

Chapter 8 Challenge Set ·································

1. Find the domain and range of each function. If the function has an inverse, find the inverse function. Otherwise, find a restricted domain on which the function does have an inverse. State this inverse function. (*Section 8.2*)

a. $f(x) = x^3 + 1$ **b.** $f(x) = \sqrt[3]{x^2}$ **c.** $f(x) = \dfrac{1}{\sqrt{x-3}} + 2$

2. Suppose $z = p + qi$ is a complex number that has the property that $z^4 = i$. (*Section 8.4*)

a. Show that each of the numbers iz, i^2z, and i^3z also has the property that its fourth power is i. Express each of these numbers in $a + bi$ form.

b. Let $p = \dfrac{1}{2}\sqrt{2 + \sqrt{2}}$, and let $q = \dfrac{1}{2}\sqrt{2 - \sqrt{2}}$. Show that $p + qi$ has the property above. (*Hint*: $z^4 = (z^2)^2$.)

3. Suppose $a + bi$ and $c + di$ are complex numbers such that both their *sum* and their *product* are real. Show, algebraically, that $c = a$ and $b = -d$. How are the two numbers $a + bi$ and $c + di$ related? (*Hint*: What you are given is that the imaginary part of the sum = 0 and the imaginary part of the product = 0.) (*Section 8.4*)

4. In a complex plane, plot the number $z = \dfrac{1}{2} + \dfrac{1}{2}i$. Convert the reciprocal of z, $\dfrac{1}{z}$, to $a + bi$ form and plot this reciprocal. Describe any geometric relationships you find between the magnitudes of these two numbers and between their angles. (*Section 8.5*)

5. *Extension* *Symmetries of a square* are operations that can be performed on a cardboard square to change its orientation. For example, the square can be rotated 90° counterclockwise (CC), or it can be flipped about a diagonal. "Multiplication" of two symmetries means performing one after the other. Complete the multiplication table below, and answer these questions: Is multiplication commutative? Is the set of symmetries a group? Can you find a subset of the set which forms a group by itself? Is there more than one such subset? (Such subsets are called *subgroups*.) Use a cardboard square with vertices labeled (differently) on both sides to help you answer these questions.

×	I	R1	R2	R3	HF	VF	D1	D2
No change = I								
90° CC Rotation = R1			R3					
180° Rotation = R2					HF			
270° CC Rotation = R3	I							
Horizontal flip = HF			VF					
Vertical flip = VF				D1				
Flip about UL to LR diag. = D1				R3				
Flip about LL to UR diag. = D2						R2		

Challenge Set 46 ·······························

FOR USE WITH SECTION 9.1

Use synthetic substitution to evaluate each polynomial for the given value of the variable.

1. $2x^3 - 3x^2 + 5x - 1$; $1 + i$

2. $-x^4 + 2x^3 - 7x^2 + x - 3$; $2 - i$

3. $x^4 + 1$; $\dfrac{\sqrt{2}}{2} + \dfrac{\sqrt{2}}{2}i$

4. $x^4 + x + \sqrt{3}$; $\sqrt{3} + i$

5. a. If each of two lines passes through the same two given points, then the lines must coincide. Suppose each of two quadratic polynomials passes through the same two given points. Show that these polynomials need not coincide by constructing a counterexample.

b. How many points in common must two quadratic polynomials have in order to guarantee that their graphs coincide, that is, that they have the same coefficients? Justify your answer.

6. a. Let $f(x) = ax^2 + bx + c$. Find the first and second differences of the polynomial for $x = 1, 2, 3$, and 4, in terms of a, b, and c. (*Hint*: The first differences are $f(2) - f(1)$, $f(3) - f(2)$, and $f(4) - f(3)$. The second difference is the difference between these differences, that is, $[f(3) - f(2)] - [f(2) - f(1)]$.)

b. Suppose 1, 2, and 7, are the values of $f(x)$ at $x = 1, 2$, and 3, respectively. Use your results from part (a) to find a, b, and c.

7. a. Repeat part (a) of Exercise 6 for the polynomial $g(x) = ax^3 + bx^2 + cx + d$, but find the first, second, third, and fourth differences. (*Hint*: Use the values 1, 2, 3, 4, 5, and 6 for x.)

b. Suppose 4, 0, 2, and 16 are the values of $g(x)$ at $x = 1, 2, 3$, and 4, respectively. Use your results from part (a) to find the values of a, b, c, and d.

★ **8.** Use the method of Exercise 6 to find a quadratic polynomial whose values at 1, 2, and 3 are the *first differences* you found in part (b) of Exercise 7.

9. a. Use a calculator to find the value of $f(x) = \dfrac{3x^3 + x^2 + 2x - 5}{x^3}$ for $x = 5$, 10, 50, and 100. What number do these values seem to approach as x gets large? Does your answer to this question change if the coefficient of x^2 in the numerator is changed?

b. Make a conjecture about the approximate value of a polynomial divided by the corresponding power of x, $\dfrac{ax^n + bx^{n+1} + \ldots + px + q}{x^n}$, when x is large.

Challenge Set 47 ···

Multiply.

1. $(x + a)^3$ **2.** $(x - a)^4$ **3.** $(ax + b)^3$

★ **4. a.** Explain why $(x - a)(x^n + ax^{n-1} + a^2x^{n-2} + \ldots + a^{n-1}x + a^n) = x^{n+1} - a^{n+1}$.

    b. Find two factors whose product equals $x^5 - 32$.

★ **5. a.** Explain why, if n is even, $(x + a)(x^n - ax^{n-1} + a^2x^{n-2} - \ldots - a^{n-1}x + an)$
      $= x^{n+1} + a^{n+1}$.

    b. Find two factors whose product equals $x^5 + 100{,}000$.

Divide.

6. $\dfrac{x^2 + 3x - 1}{x - i}$ **7.** $\dfrac{8x^3 + 2x^2 + x - 5}{2x + i}$ **8.** $\dfrac{x^3 - ix - 1}{ix + 1}$

★ **9. a.** In his method of solving an equation of the form $x^3 + bx^2 + cx + d = 0$, the 16th century mathematician Niccolo Tartaglia first wrote the equation in terms of another variable u, so that the u^2 term would have coefficient 0. Show that this can be done by setting $u = x + \dfrac{b}{3}$. Write the equation you get.

    b. If a number u is a solution of the equation you found in part (a), describe what you would do to find a solution of the original equation.

10. a. Let $z_1 = -1 + i\sqrt{3}$ and $z_2 = -1 - i\sqrt{3}$. Find the cube of each of these numbers.

    b. What is the product of the two numbers given in part (a)?

11. A *power series* $a_0 + a_1x + a_2x^2 + a_3x^3 + \ldots$ is an "infinitely long" polynomial. If a function that is difficult to evaluate can be expressed as a power series, it can be easily approximated by evaluating several terms of the power series. One such function is

$$\frac{1}{1 - x} = 1 + x + x^2 + x^3 + \ldots.$$

    a. Find a power series for $\dfrac{1}{1 + x} = \dfrac{1}{1 - (-x)}$.

    b. Multiply the series for $\dfrac{1}{1 - x}$ by the series for $\dfrac{1}{1 + x}$ to get another power series. Is your answer reasonable in view of the algebraic product of these two "fractions"? Explain.

Challenge Set 48 ·····························

1. a. Using the following examples,

$$f(x) = x^3 + 3x^2 - x + 4 \qquad f(x) = -x^3 - 5x + 10 \qquad f(x) = \frac{1}{4}x^5 - 3x^4 + x - 2$$

investigate the end behavior of odd-degree polynomial functions. Make a general statement about the end behavior of any odd-degree polynomial function.

b. Using your statement from part (a), explain how you know that any odd-degree polynomial must have a real zero.

c. Explain how you know that an odd-degree polynomial cannot have either a maximum or a minimum.

In Exercises 2 and 3:

a. Graph both functions f and g on one set of axes.

b. Describe the similarities in the graphs of the two functions near a specific point on each.

2. $f(x) = 2x^2$; $g(x) = 2x(x - 3)^2$ **3.** $f(x) = \frac{1}{5}(x + 1)^3$; $g(x) = \frac{1}{5}x(x - 4)^3$

4. *Writing* By drawing the graphs of several odd-degree polynomials, make a conjecture about the number of times such a graph can intersect the *x*-axis. Do the same investigation for even-degree polynomials. Explain your results in terms of local maximums and/or local minimums.

5. A conical flask that is part of a chemical process must fit inside a sphere of radius 10 cm, as shown. Let *x* = the distance from the base of the cone to the center of the sphere.

a. Use the Pythagorean theorem to express the radius *r* of the cone as a function of the distance *x*. What is the domain of *x*?

b. Use the result of part (a) to express the volume of the flask as a function of *x*. (*Hint*: The volume of a cone is given by the formula volume = $\frac{1}{3}\pi(\text{radius})^2(\text{height})$.)

c. Graph the function you found in part (b). Find the value of *x* that maximizes the volume of the cone. What height maximizes this volume?

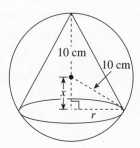

Challenge Set 49 ·····························

1. a. Graph each of the following functions.

$$f(x) = x^3 \qquad f(x) = -x^3 \qquad f(x) = -\frac{1}{2}x(x-3)^5 \qquad f(x) = 0.1(x-6)(x+5)^7$$

b. What can you conclude about the graph of a polynomial equation with an odd-numbered multiple zero near the point on the graph corresponding to the zero?

2. *Writing* Repeat Exercise 1, but use $f(x) = x^2$ and $f(x) = -x^2$, as well as several other polynomials with even-numbered multiple zeros. What can you conclude about the nature of the graph near such a multiple zero?

3. Two of the zeros of $x^4 - 3x^3 - 9x^2 - 3x - 10 = 0$ are i and $-i$. Use the Factor Theorem to find the other two zeros.

★ **4.** Let $f(x) = (x - p)(x - q)(x - r)$ be a cubic equation in factored form. By multiplying out the factors, state a theorem about the relationship between the zeros of a cubic polynomial and the constant term of the polynomial. With appropriate modification, extend your theorem to fourth-degree polynomials, and prove it for this situation. State a version of your theorem that would apply to polynomials of any degree.

5. Let $P(x) = ax^3 + bx^2 + cx + d$ be a cubic polynomial with integral coefficients a, b, c, and d, and suppose r is a zero of P that is an integer .

a. Solve the equation $P(r) = 0$ for $-d$, and show algebraically that r is a factor of the left-hand side of the resulting equation.

b. What does part (a) prove about a cubic polynomial with integral coefficients that has an integer r as a zero?

6. Suppose P is a polynomial. You know that when $P(x)$ is divided by a polynomial $d(x)$ of smaller degree than P, you can write

$$P(x) = d(x)Q(x) + R(x),$$

where Q and R are polynomials, and the degree of R is less than the degree of d.

a. Suppose $d(x)$ is of the form $x - k$ for some number a. Rewrite the equation above using this special form of d. Why is it reasonable to call the last term R in this case, rather than $R(x)$?

★ **b.** Use the equation above to prove the Factor Theorem: $x - k$ is a factor of P if and only if k is a zero of P. (*Hint*: $x - k$ is a factor of P means that $R = 0$.)

Challenge Set 50 ··

Use the Rational Zeros Theorem to prove that each number is irrational.

EXAMPLE: $\sqrt{2}$

SOLUTION: $\sqrt{2}$ is a zero of the function $f(x) = x^2 - 2$. According to the Rational Zeros Theorem, the only possible rational zeros of this function are ± 1, ± 2. Since none of these is in fact a zero of f, f cannot have any rational zeros. Therefore, $\sqrt{2}$ must be irrational.

1. $\sqrt{5}$ **2.** $\sqrt[3]{2}$ **3.** $\sqrt[5]{6}$ **4.** $\sqrt{\dfrac{2}{3}}$

5. Show that $f(x) = 2x^3 - 5x + 1$ has no rational zeros, but that f has a real zero between 1 and 2. (*Hint*: Draw the graph of f.)

★ **6.** Suppose P is a polynomial with a positive leading coefficient. It can be shown that if P is divided by $x - m$, and the coefficients of the quotient are all positive numbers, then P has no zero greater than m. Use this theorem to find the smallest positive integer m such that the given polynomial P has no zero greater than m.

a. $P(x) = x^3 - 2x^2 - 3x + 1$ **b.** $P(x) = 2x^3 - 2x^2 - 15x - 20$

7. Let $f(x) = ax^3 + bx^2 + cx + d$. The Remainder Theorem for f asserts that the remainder R that you get when you divide f by $x - k$ is equal to $f(k)$. Use the following method to prove this theorem. You can write $f(x) = (x - k)Q(x) + R$, where Q is a polynomial. Substitute $x = k$ into this equation.

8. The Rational Zeros Theorem, applied to a cubic equation, asserts that if the function $f(x) = ax^3 + bx^2 + cx + d$ with integral coefficients a, b, c, and d has a rational zero $\dfrac{p}{q}$, then p is a divisor of the constant term d, and q is a divisor of the leading coefficient a. Assume p and q have no common factor and prove this theorem using the following method.

a. Write out the equation $f\left(\dfrac{p}{q}\right) = 0$. Multiply both sides by q^3 and solve the equation for ap^3. Why does the new equation show that q must be a factor of a?

b. Again write out $f\left(\dfrac{p}{q}\right) = 0$. Multiply both sides by $\left(\dfrac{q}{p}\right)^3$.

Compare the new equation to the old one and notice that "only the names are changed." Use the argument in part (a).

Challenge Set 51

1. a. Suppose $\dfrac{1}{x} \le \dfrac{1}{y}$, for two positive numbers x and y. Prove that $x \ge y$.

  b. Suppose x and y are *negative* numbers satisfying the inequality in part (a). How are x and y related? Suppose x is negative and y is positive. How are x and y related?

Solve each inequality using the rules you discovered in Exercise 1.

2. $\dfrac{1}{x+4} \le \dfrac{1}{2x}, x > 0$ **3.** $\dfrac{2}{a+3} > \dfrac{1}{3a+4}, a > -\dfrac{4}{3}$ **4.** $\dfrac{5}{7n-1} < \dfrac{3}{4n}, n > \dfrac{1}{7}$

5. The gravitational force that Earth exerts on an object (its weight) is inversely proportional to the *square* of the object's distance from the center of Earth. The surface of Earth is about 4000 mi from its center. Suppose you weigh 85 lb on Earth.

  a. How much would you weigh at an altitude of 500 mi above the surface of Earth?

  b. How far away from the center of Earth would you have to be in order to weigh 1 lb?

6. It can be shown that the area of the region bounded by the graph of $y = \dfrac{1}{x}$,

the x-axis, and the lines $x = 1$ and $x = a$ is $\ln a$. Let A_a^b denote the area

below the graph between $x = a$ and $x = b$, $(1 \le a < b)$, so that $A_1^a = \ln a$.

In the following exercises, keep in mind that $A_a^b = A_a^k + A_k^b$, if $a \le k \le b$.

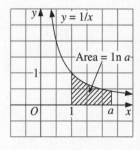

  a. Evaluate $A_1^a + A_a^{ab}$ and $A_1^a + A_1^b$ in terms of $\ln a$ and $\ln b$.

  b. State a relationship between two of the areas involved in part (a).

7. Denote by E_n the *difference* between the combined area of n rectangles like those shown and $A_1^{n+1} = \ln(n+1)$. (E_4 is illustrated in the diagram.) It can be shown that E_n approaches a fixed number, called Euler's Constant, denoted γ, as $n \to +\infty$. (It has not been shown yet whether γ is rational or irrational.) Approximate γ using 5, 10, and 15 rectangles. Make a conjecture about the value of γ.

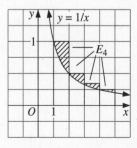

Challenge Set 52 ···

Write *y* as a rational function of *x*.

1. $xy + x - 3 = y$ 　　　　**2.** $y - 4xy + 2x = 5$ 　　　　**3.** $x + 7 = \dfrac{2-x}{y}$

4. Find, in terms of *a*, *h*, and *k,* the inverse function of $f(x) = \dfrac{a}{x-h} + k$.

5. a. The *axes* of a hyperbola whose equation has the form $y = \dfrac{a}{x-h} + k$ are the two lines through the point (h, k) with slopes 1 and –1. Find equations of these lines.

   b. Only one of the axes of a hyperbola intersects the hyperbola. How can you tell from the equation of the hyperbola which one it is? (*Hint*: Graph several examples.)

   c. Assume that the axis with slope 1 intersects the hyperbola. Find the coordinates of the point of intersection in terms of *a*, *h*, and *k*.

In Exercises 6 and 7, find an equation for the specified hyperbola in the form $y = \dfrac{a}{x-h} + k$.

6. The hyperbola has the line $y = 4$ as its horizontal asymptote and passes through the points $(-1, 3)$ and $(3, 7)$.

7. The hyperbola has the line $x = -5$ as its vertical asymptote and passes through the points $(-3, 4)$ and $(5, 0)$.

8. Graph the hyperbolas $y = \dfrac{a}{x-3} + 2$ for $a = \dfrac{1}{2}, 1, 3, 5, -2, -5$. Describe how the value of *a* affects the geometry of the hyperbola.

9. a. Graph the functions $y = \dfrac{1}{(x-2)^2} + 3$ and $y = \dfrac{1}{(x-2)^3} + 3$ in different coordinate planes.

   b. What differences and similarities do you notice between the graphs in part (a) and the hyperbola $y = \dfrac{1}{x-2} + 3$?

   c. On the basis of your observations in part (b), sketch the graph of $y = \dfrac{1}{(x-2)^4} + 3$ without using technology.

Challenge Set 53 ·······································

Graph each rational function.

1. $f(x) = \dfrac{3}{(x-2)^2}$　　　　**2.** $f(x) = \dfrac{-2}{(x+5)^3}$　　　　**3.** $f(x) = \dfrac{4}{x(x-3)^2}$

4. *Writing* Based on your answers to Exercises 1–3, describe the behavior of the graph of a rational function near a zero of the denominator, if the denominator has a factor of the form $(x-a)^n$. Cover the cases where n is even and where n is odd separately.

Rewrite each rational function $\dfrac{f(x)}{(x-a)(x-b)\ldots(x-q)}$ **in the form** $\dfrac{A}{x-a} + \dfrac{B}{x-b} + \ldots + \dfrac{Q}{x-q}$,

where A, B, . . . , Q are real numbers.

EXAMPLE: $\dfrac{2x+11}{(x-2)(x+3)}$

SOLUTION: $\dfrac{2x+11}{(x-2)(x+3)} = \dfrac{A}{x-2} + \dfrac{B}{x+3} = \dfrac{A(x+3) + B(x-2)}{(x-2)(x+3)}$

Equate the coefficients of like powers of x: $2 = A + B$
$$11 = 3A - 2B$$

The solution of this system is $A = 3$ and $B = -1$. Therefore,

$$\dfrac{2x+11}{(x-2)(x+3)} = \dfrac{3}{x-2} - \dfrac{1}{x+3}.$$

5. $\dfrac{-3x+18}{(x-1)(x-4)}$　　　　**6.** $\dfrac{7x-3}{x^2-9}$　　　　**7.** $\dfrac{4x-2}{x^2+4x+3}$

★ **8.** The *Witch of Agnesi* is a curve discovered by the Italian mathematician Maria Gaetana Agnesi (1718-1799). (The name "witch" was an English mistranslation of the Italian for "turning.") The curve has the equation

$$y = \dfrac{8}{x^2+4}.$$

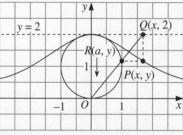

Let $P(x, y)$ be a point on the curve. Show that the points $Q(x, 2)$ directly above P on the line $y = 2$ and $R(a, y)$ directly to the left of P on the circle of radius 1 centered at $(0, 1)$ lie on the same line through the origin. Use the following method. Write y in terms of x and then use the slope formula to find a. Show that the pair (a, y) satisfies the equation of the circle $x^2 + (y-1)^2 = 1$.

Challenge Set 54 ·····························

Solve each equation.

1. $\dfrac{a}{a-3} - \dfrac{7a-6}{a^2-a-6} = \dfrac{2}{a+2}$

2. $\dfrac{x}{x+3} - \dfrac{x}{x-1} = \dfrac{8}{x^2+2x-3}$

3. $\dfrac{y+3}{y^2-2y} - \dfrac{y-10}{y^2-4} = \dfrac{y+4}{y^2+2y}$

4. $\dfrac{2b-5}{b-3} - \dfrac{3}{2b^2-7b+3} = 3$

5. $\dfrac{x-2}{x-1} + \dfrac{4}{x^2-1} = \dfrac{x-3}{x+1}$

6. $\dfrac{r-2}{r^2-1} - \dfrac{2r-1}{r^2+2r-3} = \dfrac{3}{r^2+4r+3}$

7. $\dfrac{1}{n-3} + \dfrac{n}{(n-3)^2} = \dfrac{2n+4}{(n-3)^3}$

8. $\dfrac{2x+4}{2x-1} - \dfrac{17-x}{2x^2+5x-3} = 2$

Show that each equation has no solution.

9. $\dfrac{4}{p^2-p} + \dfrac{4}{p} = \dfrac{3}{2p-2}$

10. $\dfrac{1}{u-1} + \dfrac{1}{u-2} = \dfrac{u^2-u-1}{u^2-3u+2}$

11. $\dfrac{x+4}{x^2+2x} - \dfrac{x}{x^2+3x} = \dfrac{5}{x^2+5x+6}$

12. $\dfrac{6}{x-1} - \dfrac{x}{x-3} = \dfrac{x-x^2}{x^2-4x+3}$

13. Solve the equation

$$\frac{k-2x}{k+x} + \frac{k+4x}{2k-x} = 2$$

for x, in terms of k.

14. a. Solve the equation

$$\frac{cx-d}{cx+d} = \frac{x-2d}{x+d}$$

for x in terms of c and d.

b. For what value(s) of c and/or d is there no solution of the equation?

★ **15.** An airliner made a round trip to a city 560 mi from its point of origin and encountered the same wind in both directions (a tail wind in one direction and a head wind in the other). The plane's speed relative to the surrounding air was 245 mi/h in both directions and the plane averaged 240 mi/h for the entire trip. What was the speed of the wind?

Chapter 9 Challenge Set ···

1. The mathematician Joseph Louis Lagrange (1736–1813) discovered a simple formula to solve the following problem. Given the points (x_0, y_0), (x_1, y_1), (x_2, y_2), . . . , (x_n, y_n), find a polynomial function P of degree n whose graph passes through all of these points. For $n = 2$, this formula is

$$P(x) = y_0 \bullet \frac{(x - x_1)(x - x_2)}{(x_0 - x_1)(x_0 - x_2)} + y_1 \bullet \frac{(x - x_0)(x - x_2)}{(x_1 - x_0)(x_1 - x_2)} + y_2 \bullet \frac{(x - x_0)(x - x_1)}{(x_2 - x_0)(x_2 - x_1)}.$$

a. Explain how you know $P(x)$ passes through the points (x_0, y_0), (x_1, y_1), and (x_2, y_2) and has degree 2.

b. Give Lagrange's formula for $n = 3$. (*Section 9.3*)

2. *Descartes' Rule of Signs* provides information about the possible numbers of positive and negative real zeros of a polynomial. It works this way for the polynomial $P(x) = x^5 + 2x^4 - 3x^3 + x - 5$. Count the number of *changes* in the signs of the coefficients when the polynomial is written in standard form. Here, there are three changes: $+2 \rightarrow -3$; $-3 \rightarrow +1$; $+1 \rightarrow -5$. Descartes' Rule says that for k changes, there are $k - 2n$ positive zeros for some positive integer n or 0. You can also apply the rule to $P(-x)$ to find the possible numbers of negative zeros. (Here, $P(-x) = -x^5 + 2x^4 + 3x^3 - x - 5$. There are two changes of sign.) Again, for k changes there are $k - 2n$ negative zeros.
(n may be 0.) For each polynomial make a table listing the possibilities for the numbers of positive, negative, and imaginary zeros, using one row for each distinct possibility. (*Section 9.5*)

a. $P(x) = x^5 - 3x^4 + 7x^2 - 6x - 8$ **b.** $P(x) = x^6 + x^5 - 3x^3 - 2x^2 + 5x - 10$

3. Simplify each expression. (*Section 9.8*)

a. $\left(\dfrac{x}{y} - \dfrac{y}{x} \right) \div \left(\dfrac{x^2}{y^2} - 1 \right)$ **b.** $\left(a - b + \dfrac{a^2}{a + b} \right) \div \left(a + b + \dfrac{a^2}{a - b} \right)$

4. *Extension* Suppose you want to find a zero of the polynomial $P(x) = 2x^3 - 3x^2 - x - 1$. Since $P(1) = -3$ and $P(2) = 1$, you know that the graph of $P(x) = 0$ must cross the x-axis between $x = 1$ and $x = 2$. Now, $P(1.5) = -2.5$. Therefore, the crossing occurs between $x = 1.5$ and $x = 2$. Continuing in this way, you can narrow the interval in which the graph crosses the x-axis (and thus in which P has a zero) indefinitely. Use this method to locate a zero of each polynomial in the specified domain to within $\pm\dfrac{1}{32}$.

a. $P(x) = x^3 - x^2 - 4x - 2$; $2 \le x \le 3$ **b.** $P(x) = -x^3 + 2x^2 - 3x + 1$; $0 \le x \le 1$

Challenge Set 55 ···

FOR USE WITH SECTION 10.1

Write the next four terms in each sequence.

1. 2, 5, 10, 17, 26, . . .

2. 256, 16, 4, 2, . . .

3. O, T, T, F, F, . . .
(*Hint*: Think of integers.)

4. $1, \frac{1}{2}, 3, \frac{1}{4}, 5, \frac{1}{6}, \ldots$

For each sequence, write a formula for t_n, and find the next three terms.

5. $\frac{1}{2}, \frac{4}{3}, \frac{3}{4}, \frac{6}{5}, \ldots$ (*Hint*: Use $(-1)^n$.)

6. 7.7, 7.07, 7.007, 7.0007

7. Draw the next figure in the sequence shown.

, . . .

8. The maximum numbers of electrons that can be contained in each spherical layer, or *shell*, of electrons in an atom are 2, 8, 18, 32, ... Find the next term in this sequence and write a formula for t_n.

9. The numbers of diagonals of n-sided polygons for $n = 3, 4, 5, 6, \ldots$ are 0, 2, 5, 9, . . . Find the next two numbers in this sequence. Draw diagrams of the relevant polygons and count the diagonals to check your answers.

10. a. Consider the sequence of cubes of the nonnegative integers: 0, 1, 8, 27, ... Write the first five terms of the sequence of differences, $t_n - t_{n-1}$, between terms of this sequence.

b. Write the first five terms of the sequence of *hexagonal numbers*; that is, those numbers that represent the numbers of dots in the sequence of diagrams like the following.

c. What relationship do you notice between your answers to parts (a) and (b)? Account for this relationship using the geometric representation of a perfect cube suggested by the diagram at the right.

Challenge Set 56 ···

1. If a and b are two given positive numbers, with $a \leq b$, inserting *two* geometric means between a and b means finding x and y such that the sequence a, x, y, b is geometric. Similarly, you could insert three or more geometric means. For each sequence, insert the specified number of geometric means between the two given numbers.

    a. two between -2 and 128 **b.** three between 8 and 40.5

    c. three between 3 and 12 **d.** four between $\dfrac{2}{25}$ and $\dfrac{125}{16}$

2. If a and b are any two positive numbers, with $a \leq b$, prove that their arithmetic mean is greater than or equal to their geometric mean. (*Hint*: Start with the fact that $(a - b)^2 \geq 0$, for any a and b. Add $4ab$ to both sides, and take the square root of both sides.) Can it happen that these two means are equal? If so, give an example.

★ **3.** In a right triangle, such as triangle ABC at the right, the altitude $\overline{CD}$ to the hypotenuse is the geometric mean between the two segments that it divides the hypotenuse into: $\overline{AD}$ and $\overline{BD}$. Also, the *median* $\overline{CE}$ drawn to the hypotenuse is half the length of the hypotenuse. Use these facts to give an alternative proof of the fact that the arithmetic mean of two positive numbers is greater than or equal to their geometric mean.

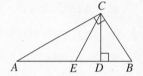

4. Suppose the sequence $t_1, t_2, t_3, \ldots$ is geometric. Does the sequence $\log t_1, \log t_2, \log t_3, \ldots$ have any special properties? What kind of sequence is it? Justify your answer.

5. Suppose the sequence $a_1, a_2, a_3, \ldots$ is arithmetic. What kind of sequence is the sequence $2^{a_1}, 2^{a_2}, 2^{a_3}, \ldots$? Justify your answer.

★ **6.** Suppose that the sequence a, b, c is *both* arithmetic and geometric. Prove that $a, b,$ and c must all be equal. (*Hint*: Let $b = a + x$. Write c in terms of a and x. Use the fact that b is the geometric mean between a and c to show that x must be 0.)

7. Find an arithmetic sequence none of whose terms is divisible by 2, 3, or 5. (*Hint*: Find an arithmetic sequence all of whose terms are divisible by all these numbers. Then modify it slightly.)

8. Find x so that the sequence $x - 2, x, 2x + 3$ is:

    a. arithmetic   **b.** geometric

Challenge Set 57 ·····································

FOR USE WITH SECTION 10.3

Write a recursive definition for each sequence.

1. $a, a^2 + a, a^3 + a^2 + a, \ldots$ **2.** $a, ab - a, ab^2 - ab - a, \ldots$

3. The number $n!$, read as "n factorial," is defined to be $1 \cdot 2 \cdot 3 \ldots \cdot n$. Define the sequence $1!, 2!, 3!, \ldots$ recursively.

4. Suppose you define a sequence recursively by the equations

$$t_1 = 1; \, t_n = t_{n-1} + 2\sqrt{t_{n-1}} + 1.$$

    a. Write out the first five terms of the sequence.

    b. Write an explicit formula for the terms of the sequence.

★   **c.** Assume that your formula is true for t_{k-1} and use the recursion equation to show that it is true for t_k.

5. *Rectangular numbers* can be represented geometrically by the numbers of dots in the diagrams at the right.

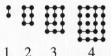

1   2   3   4

    a. Find the first five rectangular numbers.

    b. Write a recursive formula for the sequence of rectangular numbers.

    c. Write an explicit formula for the sequence. (*Hint*: Notice that each diagram at the right contains a square.)

6. The *Sierpinski Curve* is a curve that theoretically requires infinitely many stages to complete. The first three stages are shown schematically below. (The curve itself is the *outline* of the design and should be scaled so that all stages are the same size.)

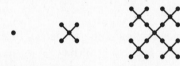

Stage 0   Stage 1   Stage 2

    a. Sketch Stage 3 of the Sierpinski Curve schematically. (*Hint*: Each stage attaches four copies of the preceding stage around one central dot.) Write the first four terms of the sequence of the numbers of dots in the stages of the schematic Sierpinski Curve.

    b. Give a recursive formula for the sequence in part (a).

Challenge Set 58 ···

1. Find the sum of all the odd integers between 100 and 200.

2. a. Find a formula, in terms of n, for the sum of the first n even integers.

b. Use the formula you found in part (a) to calculate the sum of all the even integers between 1 and 99.

c. Suppose the sum of all the even integers from 2 through x is 240. Find the even number x.

★ **d.** Generalize the formula you found in part (a) so that it gives the sum of the first n multiples of any positive integer p in terms of n and p.

Find the sum of each arithmetic series in terms of a and b.

3. $(a + b) + (a + 2b) + (a + 3b) + \ldots + (a + 12b)$

4. $\left(a - \dfrac{b}{2}\right) + \left(a - \dfrac{3b}{2}\right) + \left(a - \dfrac{5b}{2}\right) + \ldots + \left(a - \dfrac{23b}{2}\right)$

5. a. Find a formula for the sum of the first n positive integers from the general formula for the sum of an arithmetic series.

b. Another way of proving the formula for the sum of the first n positive integers is called *mathematical induction*. In this method, you prove that the formula works for $n = 1$, and you also prove that *if* it works for a *particular* value $n = k$, then it also works for $n = k + 1$. These two facts, taken together, imply that the formula works for all positive integers n. Verify the formula in part (a) for $n = 1$. Then assume that the formula is true for $n = k$,

$$1 + 2 + 3 + \ldots + k = \text{(the expression for the sum of } k \text{ terms),}$$

and prove that the formula works for $n = k + 1$ by adding $k + 1$ to both sides.

★ **6.** A formula for the sum of the *squares* of the first n positive integers is

$$1^2 + 2^2 + 3^2 + \ldots + n^2 = \frac{n(n + 1)(2n + 1)}{6}.$$

Prove this formula using mathematical induction. (*Hint*: See Exercise 5.)

7. Use the formula in Exercise 6 to find the sum of the squares of all positive integers between 10 and 30, inclusive. (*Hint*: Apply the formula with $n = 30$ and $n = 9$.)

Challenge Set 59 ·······································

1. a. Show that $\dfrac{1}{2} + \dfrac{1}{4} + \dfrac{1}{8} + \ldots + \dfrac{1}{2^n} = 1 - \dfrac{1}{2^n}$.

★ **b.** Is it true in general that if p is a positive integer, $\dfrac{1}{p} + \dfrac{1}{p^2} + \dfrac{1}{p^3} + \ldots +$

$\dfrac{1}{p^n} = 1 - \dfrac{1}{p^n}$? If so, prove this. If not, give a correct formula for the

left side of the expression.

2. a. Simplify $(1 - x)(1 + x + x^2 + \ldots + x^{n-1})$.

★ **b.** Use the result of part (a) to prove the formula for the sum of a geometric series

$$a + ar + ar^2 + \ldots + ar^{n-1} = \dfrac{a(1 - r^n)}{1 - r}.$$

Find the sum of each series in terms of *a* or *b*.

3. $b + b(b + 1) + b(b + 1)^2 + b(b + 1)^3 + \ldots + b(b + 1)^n$

4. $\dfrac{1}{a} + \dfrac{a - 1}{a^2} + \dfrac{(a - 1)^2}{a^3} + \ldots + \dfrac{(a - 1)^{n-1}}{a^n}$

5. In any number system, the fractional part of a number is written as a sum of digits times negative powers of the base, just as in the decimal system. For example, in base 3, the notation 0.1201 means $1 \cdot 3^{-1} + 2 \cdot 3^{-2} + 0 \cdot 3^{-4} +$

$1 \cdot 3^{-5}$, or $\dfrac{1}{3} + \dfrac{2}{9} + \dfrac{0}{81} + \dfrac{1}{243}$. Write each nonterminating "decimal" as an

ordinary fraction in the number system having the specified base.

a. 0.1010101 . . . ; base 2   **b.** 0.333333 . . . ; base 5

c. 0.2020202 . . . ; base 3   **d.** 0.313131. . . ; base 4

6. Suppose you start with a square, drop out the center smaller square as shown below, and continue in this way dropping the center squares of each of the squares thus created.

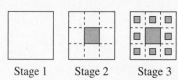

Stage 1   Stage 2   Stage 3

a. Suppose the original square has sides of length 1. Write, as a series, the combined areas of the smaller squares that are dropped out at each stage.

b. What is the total area of the shaded parts that are dropped after infinitely many stages?

NAME _____ DATE _____

Chapter 10 Challenge Set ·······························

1. a. The diagram at the right shows that 7 is the maximum number of regions into which 3 lines divide a circle. Write out the first five terms of the sequence of numbers of regions t_n that n lines divide a circle into.

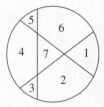

    b. Write a recursive formula for the sequence in part (a). Observe that when you add a line to the diagram illustrating t_{n-1}, it intersects a maximum of $n - 1$ other lines. (*Section 10.3*)

2. a. Find the sums $\sum_{k=1}^{n} \dfrac{1}{k(k+1)}$ for $n = 1, 2, 3,$ and 4.

    b. Make a conjecture about the partial sum $\sum_{k=1}^{n} \dfrac{1}{k(k+1)}$ in terms of n.

    Prove your conjecture. $\left(Hint: \dfrac{1}{k(k+1)} = \dfrac{1}{k} - \dfrac{1}{k+1}.\right)$

    c. Why do you think a sum like that in part (b) is called a "telescoping sum"? (*Section 10.4*)

3. In this exercise, you will find a formula for the sum $1^3 + 2^3 + 3^3 + \ldots + n^3$. Each geometric cube in the first diagram at the right represents one of the cubes in the sum. In the second diagram these cubes are taken apart and arranged in a single layer of small cubes in the shape of a square.

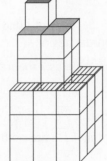

    a. Show how to arrange the next larger cube ($4 \times 4 \times 4$, not shown in the diagram) similarly in the single-layer arrangement.

    b. Give a simple formula, in terms of n, for the number of small cubes *along one edge* of the single-layer square, after n large cubes have been taken apart and arranged.

    c. Based on these diagrams and the result of part (b), find a formula for the sum of the first n cubes. (*Section 10.4*)

4. *Extension* Suppose z and r are *complex numbers*, and you form the geometric sequence $a, ar, ar^2, ar^3, \ldots$. What pattern do the points in the sequence make when graphed in the complex plane? To answer this question, use $a = 1$ and the following three values of r: $1 + 0.6i$, $0.6 + 0.8i$, and $0.5 + 0.5i$. In different complex planes, graph the three sequences that you obtain from these values. (*Hint:*Use large units.) How is the pattern of the dots related to the magnitude of r? Try using the same values of r with a different value of a. Does the same relationship hold? What changes if you replace the given r-values with their conjugates?

Challenge Set 60 ···

1. Let $P(0, 0)$ and $Q(2a, 2)$ be the endpoints of a segment $\overline{PQ}$.

   a. Find the equation of the perpendicular bisector of $\overline{PQ}$.

   b. Show that any point (x, y) on this perpendicular bisector is the same distance from P as it is from Q. (*Hint*: Replace y by the right-hand side of the equation you found in part (a).)

2. Let $A(a, 0)$, $B(-a, 0)$, and $C(b, c)$ be the vertices of a triangle, as shown.

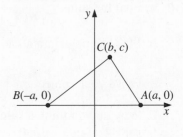

   a. Write and simplify an equation that states that the product of the slopes of $\overline{AC}$ and $\overline{BC}$ is -1.

   b. Write and simplify an equation that states that the sides of $\triangle ABC$ satisfy the Pythagorean theorem; that is, that $\triangle ABC$ is a right triangle.

   c. Show that the equations you found in parts (a) and (b) are equivalent. What does this prove?

3. a. M, N, P, and Q are the midpoints of the sides of quadrilateral $ABCD$, as shown in the diagram. Find their coordinates of M, N, P, and Q.

   b. Show that $MNPQ$ is a parallelogram. State as a theorem the fact you have proved.

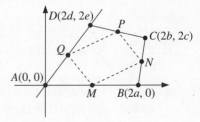

4. a. Show that, in the diagram for Exercise 3, $\overline{PQ}$ and $\overline{MN}$ are both parallel to diagonal $\overline{AC}$, and that $\overline{QM}$ and $\overline{PN}$ are both parallel to diagonal $\overline{BD}$.

   b. Suppose diagonals $\overline{AC}$ and $\overline{BD}$ are perpendicular. What special kind of quadrilateral is $MNPQ$?

5. Let l be the line $y = x$, let $P(x, x)$ be any point on this line, and let $Q(a, b)$ be any point in the coordinate plane.

   a. Find x so that the distance PQ is minimized. (*Hint*: Write this distance in terms of a, b, and x. Note that a square root is minimized when the number under the radical sign is minimized, and note also that the expression under the radical sign is a quadratic function of x.)

   b. Show that with the value of x that you found in part (a), $\overline{PQ}$ is perpendicular to l.

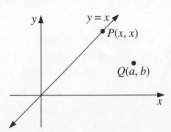

Challenge Set 61 ···

Find the points of intersection of each parabola and line.

1. $y = 2x^2 - x + 1$

$y = -3x + 5$

2. $y = -\frac{1}{2}x^2 + 3x + 1$

$y = -2x + 13$

3. $y = 3x^2 + 2x - 4$

$y = 8x + 5$

Find the point(s) of intersection of each pair of parabolas.

4. $y = x^2 + 2x - 5$

$y = -x^2 - x + 9$

5. $y = 2x^2 - 5x - 1$

$y = x^2 + 3x + 8$

6. $y = \frac{2}{3}x^2 + 4x - 8$

$y = 2x^2 - 4x + 1$

7. You can find the slope of the *tangent line* to a parabola $y = x^2$ at a point $P(a, a^2)$ using the following method.

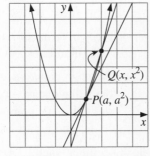

    a. Find an expression, in terms of x and a, for the slope of a *secant line* from P to another point $Q(x, x^2)$ on the parabola.

    b. Simplify this slope, assuming $x \neq a$. What number, in terms of a, does this quantity approach as x approaches a?

    c. Explain geometrically why the procedure in part (b) gives you the slope of the tangent line.

★ **8.** In the diagram at the right, F is the focus of the parabola $y = x^2$, and the line d is its directrix. The line l is tangent to the parabola at the point $P(a, a^2)$, $\overline{PF}$ is drawn, and $\overline{PD}$ is drawn perpendicular to d.

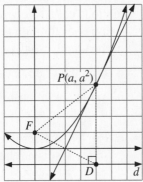

    a. Find the coordinates of F and D.

    b. Use the slope you found in part (b) of Exercise 7 to find the equation of line l.

    c. Find an equation of the line $\overleftrightarrow{FD}$.

    d. Show that line l is the perpendicular bisector of $\overline{FD}$ by showing that the lines are perpendicular and that the midpoint of $\overline{FD}$ lies on l. What does this tell you about the relationship of line l to $\angle FPD$?

Challenge Set 62 ··

FOR USE WITH SECTION 11.3

In Exercises 1–3, an equation of a circle and an equation of a parabola are given. Find all the points of intersection of the two graphs.

1. $y = x^2 - 13$

    $x^2 + y^2 = 25$

2. $y = -\frac{1}{2}x^2 + 17$

    $x^2 + y^2 = 289$

3. $x = \frac{1}{5}y^2$

    $(x + 7)^2 + y^2 = 169$

Graph each inequality.

4. $x^2 + y^2 < 16$

5. $(x - 2)^2 + y^2 \geq 4$

6. $(x + 3)^2 + (y - 5)^2 \leq 9$

7. A ring on an archery target is bounded by two circles of radii 2 ft and 3 ft, respectively. Suppose the origin is at the center of the target. Write a system of inequalities whose solution is the ring.

8. a. Find, in terms of x and y, expressions for the distance between $A(0, 0)$ and $P(x, y)$ and the distance between $B(6, 0)$ and $P(x, y)$.

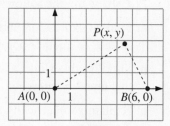

    b. Write an equation stating that the distance between A and P is twice the distance between B and P. Simplify this equation and show that the points satisfying it all lie on a circle. What are the center and radius of this circle?

9. A segment, such as $\overline{AB}$ in the diagram, that joins two points on a circle is called a *chord*. This exercise will show that a radius of a circle that passes through the midpoint of a chord that is not a diameter is perpendicular to the chord.

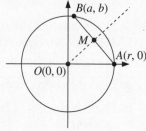

    a. Find the midpoint M of $\overline{AB}$ and the equation of $\overline{OM}$.

    b. Write a relationship between a, b, and r, based on the fact that $B(a, b)$ is on the circle that also passes through $A(r, 0)$.

★ **c.** Prove the assertion stated above, namely that $\overline{OM}$ is perpendicular to $\overline{AB}$. *(Hint*: Use the relationship you found in part (b).)

Find the equation, in standard form, of the circle passing through the three given points.

10. $(-1, 1), (5, -7), (2, 2)$

11. $(1, -7), (8, 0), (9, 5)$

Challenge Set 63 ···

FOR USE WITH SECTION 11.4

In Exercises 1–3, an equation of an ellipse and an equation of a line are given.
Find the point(s) of intersection of the two graphs.

1. $\dfrac{x^2}{2} + \dfrac{y^2}{9} = 1$

$y = -3x - 3$

2. $\dfrac{x^2}{4} + \dfrac{y^2}{12} = 1$

$y = 3x - 6$

3. $\dfrac{x^2}{45} + \dfrac{y^2}{36} = 1$

$y = \dfrac{2}{5}x + 6$

4. Find the intersection points of the ellipses

$$9x^2 + 16y^2 = 144$$

$$x^2 + 4y^2 = 26.$$

5. An ellipse can be determined by means of a focus and a directrix.
In the diagram, the directrix of an ellipse is the line
$x = \dfrac{25}{3}$, and one focus is $(3, 0)$. The relationship determining the
ellipse is $PF = \dfrac{3}{5}PD$ for all points P.

a. Use the distance formula to write an expression for PF, and
set it equal to $\dfrac{3}{5}PD$. (*Hint*: You should not need the distance

formula for PD.)

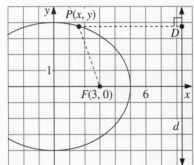

b. Simplify the equation you wrote in part (a), and put it in the
standard form of an equation of an ellipse. What is the length of the
minor axis of this ellipse?

6. For an ellipse given in standard form, the quantity $e = \dfrac{c}{a}$, where

$c = \sqrt{a^2 - b^2}$, is called the *eccentricity* of the ellipse. Graph each of the
following ellipses, and compute the eccentricity e of each.

a. $\dfrac{x^2}{25} + \dfrac{y^2}{16} = 1$

b. $\dfrac{(x - 2)^2}{26} + \dfrac{y^2}{169} = 1$

c. $\dfrac{x^2}{625} + \dfrac{(y - 4)^2}{676} = 1$

d. Based on your results in parts (a)–(c), describe the relationship between
the eccentricity of an ellipse and its geometric characteristics.

7. a. *Writing* Using the concept of eccentricity from Exercise 6, describe what
happens to the numbers a and b in the standard form of the equation of an
ellipse as e approaches 0.

b. It can be shown that an ellipse with major axis of length $2a$ and minor
axis of length $2b$ has area πab. Explain how this fact is consistent with
your answer to part (a).

Challenge Set 64 ....

FOR USE WITH SECTION 11.5

Write each equation of a hyperbola in standard form and find the coordinates of the foci.

1. $x^2 - 4y^2 - 10x - 24y - 15 = 0$ **2.** $9y^2 - 4x^2 + 16x - 90y + 173 = 0$

Graph each inequality.

3. $x^2 - y^2 < 1$ **4.** $\dfrac{y^2}{9} - x^2 \geq 1$ **5.** $xy < 4$

6. a. Graph the hyperbolas $x^2 - y^2 = k$, for $k = \pm 1, \pm 2, \pm 5$.

    b. What characteristics do all these hyperbolas share? How do you think the shape of a hyperbola of this form changes as $|k|$ approaches $+\infty$?

7. a. In one coordinate plane, graph the following four curves. Use the same size units on the x-axis and the y-axis.

    Ellipses: $\dfrac{x^2}{169} + \dfrac{y^2}{144} = 1$ Hyperbolas: $\dfrac{x^2}{16} - \dfrac{y^2}{9} = 1$

                 $\dfrac{x^2}{(6.25)^2} + \dfrac{y^2}{(3.75)^2} = 1$ $\dfrac{x^2}{9} - \dfrac{y^2}{16} = 1$

    b. What characteristic of both ellipses and hyperbolas is shared by all four of these curves? What can you say about the angles formed at the intersection points of each ellipse and each hyperbola?

8. A hyperbola can defined as a set of points the difference of whose distances from two fixed points, the foci, is a positive or negative constant. Suppose the two foci of a hyperbola are $F_1(-5, 1)$ and $F_2(5, 0)$ and the constant is 6.

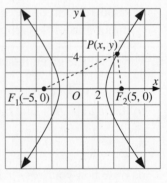

    a. Let $P(x, y)$ be a point on the hyperbola. Using the distance formula, rewrite the equation $PF_1 - PF_2 = 6$, in terms of x and y.

    b. Simplify the equation you wrote in part (a), using the following steps: (1) Get the two radicals on opposite sides of the equation and square both sides. (2) Simplify and again isolate the radical on one side of the equation. Then (3) square both sides again.

    c. Write the equation you got in part (b) in the standard form of an equation of a hyperbola.

Challenge Set 65 ··

1. The general equation of a conic $Ax^2 + Bxy + Cy^2 + Dx + Ey + F = 0$ can tell you the kind of conic it defines even when $B \neq 0$. The following rule applies: Let $K = B^2 - 4AC$. Then the equation defines an ellipse, a parabola, or a hyperbola if K is less than, equal to, or greater than 0, respectively. Tell what kind of conic each equation represents. Graph each function by solving for y. (*Hint*: To do this, you may need the quadratic formula, regarding y as the variable and x as part of the coefficient of y.)

   a. $2x^2 - xy - 10 = 0$ **b.** $x^2 + xy + y^2 - 8 = 0$ **c.** $x^2 + 2xy + y^2 - 2x = 0$

2. **Writing** Based on your answers to Exercise 1, what effect would you say the xy-term has on the graph? How do graphs containing an xy-term differ from graphs lacking such a term?

3. The hyperbola $xy = 1$ can be rotated 45° clockwise to produce a congruent hyperbola whose axes are the x- and y-axes. The equation of this new hyperbola is

$$\frac{x^2}{2} - \frac{y^2}{2} = 1.$$

   a. Graph these two hyperbolas in different coordinate systems.

   b. Use the fact that the hyperbolas are congruent and the second is a 45° clockwise rotation of the first to find the vertices and the foci of the hyperbola $xy = 1$.

★ 4. Let $A(x - h)^2 + B(y - k)^2 = C$ represent a conic.

   a. For what values of A, B, and C will the equation represent a nondegenerate ellipse? For what values will it represent a nondegenerate hyperbola?

   b. Show that a point $(h + p, y)$ is on the graph of the equation if and only if the point $(h - p, y)$ is on the graph. Describe the geometric significance of this statement.

   c. Show that a point $(x, k + q)$ is on the graph of the equation if and only if the point $(x, k - q)$ is on the graph. Describe the geometric significance of this statement.

5. The graph of $x^2 + 3xy + 2y^2 = 0$ is a degenerate conic.

   a. According to the criterion in Exercise 1, which kind of conic is it?

   b. Show that the conic is degenerate by rewriting the equation with the left side factored.

   c. Graph the conic and describe it as the intersection of a plane and a double cone.

Chapter 11 Challenge Set ·····································

1. a. What is the y-coordinate of the focus of the parabola whose equation

is $y - k = \dfrac{1}{4c}(x - h)^2$?

b. Show that the width of the parabola at this y-coordinate is $4c$.
(*Section 11.2*)

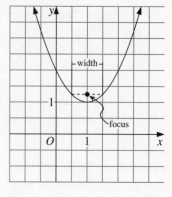

2. a. Suppose the ellipse $\dfrac{x^2}{a^2} + \dfrac{y^2}{b^2} = 1$ has foci $(c, 0)$ and $(-c, 0)$. Find

expressions, in terms of a and b only, for the y-coordinates of the
two points where the line $x = c$ intersects the ellipse. (*Section 11.4*)

b. Show that the distance between these two points is $\dfrac{2b^2}{a}$.

c. Suppose the $\dfrac{x^2}{a^2} - \dfrac{y^2}{b^2} = 1$ has foci $(c, 0)$ and $(-c, 0)$. Find an expression, in

terms of a and b only, for the distance between the two points where the
line $x = c$ intersects the hyperbola. How does this expression compare
with the result of part (b)? (*Section 11.5*)

3. Find an equation of the ellipse with one focus at the origin and the other
focus at $(10, 0)$, and with major axis of length 12. (*Section 11.4*)

4. The points $P(a, b)$ and $Q\left(\dfrac{a}{a^2 + b^2}, \dfrac{b}{a^2 + b^2}\right)$ are called *inversions*

of each other in the unit circle.

a. Show that if $O = (0, 0)$, then $OP \cdot OQ = 1$.

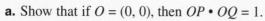

★ **b.** The equation $A(x^2 + y^2) + Dx + Ey + F = 0$ represents a
circle if A and F have opposite signs and neither is 0. Show
that the point P is on this circle if and only if the point Q is
on the circle $F(x^2 + y^2) + Dx + Ey + A = 0$. (*Section 11.6*)

5. *Extension* The hyperbola $xy = 2$ has vertices $(-\sqrt{2}, -\sqrt{2})$ and
$(\sqrt{2}, \sqrt{2})$ and foci $F_1(2, 2)$ and $F_2(-2, -2)$. Let $P(x, y)$ be a point on the
hyperbola. Using the distance formula, derive the equation $xy = 2$ from
one of the defining equations of the hyperbola $PF_1 - PF_2 = 4$.

Challenge Set 66 ···

1. a. The names of some of the suburbs of Boston, Massachusetts are shown in
the graph below. Create a possible map of these suburbs based on the graph.

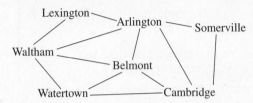

b. Color the map with 4 colors so that no two suburbs of the same color are touching.

c. Is it possible to color the map with fewer than 4 colors? If so, show how.
If not, explain why not.

2. An *incidence matrix* for a graph is a matrix with entries that are 1 if the
corresponding row element and column element are connected in the graph and 0
if they are not connected. (Assume that an element is not connected to itself.) For
example, a graph and the incidence matrix for the graph are shown below.

$$
\begin{array}{c}
 \\
A \\
B \\
C \\
D
\end{array}
\begin{array}{c}
A\ B\ C\ D \\
\left[
\begin{array}{cccc}
0 & 1 & 1 & 1 \\
1 & 0 & 0 & 1 \\
1 & 0 & 0 & 1 \\
1 & 1 & 1 & 0
\end{array}
\right]
\end{array}
$$

a. Make an incidence matrix for the each of the graphs below.

b. Make an incidence matrix for the suburbs in Exercise 1.

c. A matrix is *symmetric* if an entry and its reflection across the main
diagonal (upper left to lower right) are always equal. Explain why an
incidence matrix is always symmetric.

3. Make a graph from the incidence matrix below.

$$
\begin{array}{c}
 \\
A \\
B \\
C \\
D \\
E \\
F
\end{array}
\begin{array}{c}
A\ B\ C\ D\ E\ F \\
\left[
\begin{array}{cccccc}
0 & 1 & 0 & 1 & 1 & 0 \\
1 & 0 & 1 & 0 & 0 & 1 \\
0 & 1 & 0 & 1 & 0 & 1 \\
1 & 0 & 1 & 0 & 1 & 0 \\
1 & 0 & 0 & 1 & 0 & 1 \\
0 & 1 & 1 & 0 & 1 & 0
\end{array}
\right]
\end{array}
$$

Challenge Problems, ALGEBRA 2: EXPLORATIONS AND APPLICATIONS

Challenge Set 67 ··

1. a. Draw a directed graph representing an airline route map among five cities, A, B, C, D, and E, using the matrix at the right.

$$\begin{array}{c} \\ A \\ B \\ C \\ D \\ E \end{array} \begin{array}{c} A\ B\ C\ D\ E \\ \begin{bmatrix} 0 & 1 & 0 & 1 & 0 \\ 0 & 0 & 1 & 1 & 0 \\ 0 & 1 & 0 & 1 & 0 \\ 1 & 0 & 0 & 0 & 1 \\ 0 & 0 & 1 & 0 & 0 \end{bmatrix} \end{array}$$

b. Find a matrix that tells the number of ways of traveling from each city to each other city using 1, 2, or 3 intermediate stops.

2. Suppose F_n and S_n represent the populations of two towns, Fairview and Selbyville, n years from now. Suppose, also, that 15% of the Fairview population moves to Selbyville each year, while 10% of the Selbyville population moves to Fairview in the same time period. We then have the following relationships.

$$\left. \begin{array}{l} F_{n+1} = 0.85F_n + 0.10S_n \\ \\ S_{n+1} = 0.15F_n + 0.90S \end{array} \right\} \text{ which we can write } \begin{bmatrix} F_{n+1} \\ S_{n+1} \end{bmatrix} = \begin{bmatrix} 0.85 & 0.10 \\ 0.15 & 0.90 \end{bmatrix} \begin{bmatrix} F_n \\ S_n \end{bmatrix}.$$

a. To find out what happens to the populations of the two towns, let

$A = \begin{bmatrix} 0.85 & 0.10 \\ 0.15 & 0.90 \end{bmatrix}$. Suppose the present populations of the towns, F_0 and S_0,

are 40,000 and 10,000, respectively. Find F_1, S_1, F_2, and S_2.

b. Find A^{20} and A^{40} and use them to find the populations F_{20}, S_{20}, F_{40}, and S_{40}. (Note that you can raise a matrix to an integral power on your calculator as you would a number.)

c. Describe the long-term trend in the populations of the two towns. Would your answer change if the present populations of the two towns were reversed?

★ **3.** Populations of coyotes and rabbits share a common habitat. Each year the new coyote population is 40% of what it was the previous year plus 30% of the previous rabbit population. Each year the rabbit population becomes 120% of what it was the previous year minus 40% of the previous coyote population. (The coyotes prey on the rabbits.)

a. Write a system of equations and a *transition* matrix A, as in Exercise 2, to express the change in the populations, C_n and R_n, of coyotes and rabbits.

b. Suppose the initial populations are C_0 and R_0. Express the populations, C_{40} and R_{40}, after 40 years in terms of C_0 and R_0, using A^{40}. Describe the long-term trend.

c. Suppose that the changes in coyote and rabbit populations are as given above except that the "minus 40%" is replaced by "minus 50%." Answer the question in part (b). How would you describe what happens to the two populations?

Challenge Set 68 ··

1. a. A public-opinion survey asked respondents to rate 5 brands of instant coffee and 6 brands of cereal in order of their preference. In how many different ways can the survey questionnaire be answered?

   b. Suppose a third category consisting of 4 brands of dishwashing liquid is added to the survey. In how many ways can the questionnaire be answered?

2. a. How many 7-digit phone numbers can be formed if the first digit cannot be a 0 or a 1 and the second digit cannot be a 0?

   b. In addition to the restriction in part (a), suppose the first three digits of a phone number cannot be 411, 555, 911, or 637. How many such phone numbers are possible? (*Hint*: Use subtraction.)

3. a. License plate numbers in a certain state can have 3 letters of the alphabet followed by 1, 2, or 3 digits. How many different such license plate numbers are possible?

   b. How many different license plate numbers satisfy the conditions of part (a) and contain 3 *different* letters?

★ **4.** Part of a company logo is to have designs like the one shown, with 4 different-colored squares in the top row and 4 different-colored squares in the bottom row. The colors are all to be chosen from among 7 colors, and a further restriction is that no color in the bottom row is to match the color directly above it in the top row. How many different patterns satisfying these conditions are possible?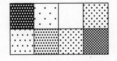

5. a. A short-answer test has 7 multiple-choice questions, each with 4 choices, followed by 5 true-false questions. In how many different ways is it possible to answer all the questions on the test?

   b. In how many different ways could you answer the multiple-choice test if you left exactly one of the true-false questions blank?

6. The baseball World Series is a best-of-seven series, which means that the first team to win 4 games wins the series. Thus, the series may comprise fewer than 7 games. In how many different ways, that is, different orders in which games are won, can series games turn out so that the American League team wins? In how many different ways altogether can the games turn out with either team winning? (*Hint*: Use a tree diagram.)

7. A public park has a pattern of paths connecting statues at *A*, *B*, *C*, and *D*, as shown. In how many ways is it possible to choose a path that visits each statue exactly once? (*Hint*: This means choosing an order for the letters *ABCD*, but taking into account that there are two ways to get from *C* to *D* and vice versa.)

Challenge Set 69 ···

FOR USE WITH SECTION 12.4

1. There are 4 teams in the Eastern Division of a basketball league, with 8 players on each team . An all-star team is to be made up of 3 representatives from each team. In how many ways can the all-star team be selected?

2. a. Fifteen factory workers are to be split up into 3 shifts, a first, a second, and a third, to maintain an assembly line. There will be 5 workers in each shift. In how many ways can the workers be assigned to the shifts?

    b. Suppose the 15 factory workers are simply to be divided into three equal groups, without assigning the groups to shifts. In how many ways can this be done? (*Hint*: Notice that in part (a), assigning workers 1–5, 6–10, and 11–15 to the first, second, and third shifts was different from assigning these groups to the third, second, and first shifts. Now these two possibilities are to be counted as one.)

3. a. In how many ways can 2 consonants and 3 vowels be chosen from among the letters of the word EDUCATION?

    b. How many sequences of 5 letters containing exactly 2 consonants and 3 vowels can be made from the letters of the word EDUCATION?

In Exercises 4–6, find the number of 5-card poker hands of each type.

4. A flush (5 cards of one suit: clubs, diamonds, hearts, or spades)

5. Three of a kind (3 cards of one denomination, such as 3 kings, and two other cards of different denominations, neither of which is of the same denomination as the first 3 cards)

6. Two pairs (two cards of one denomination, two cards of a different denomination, and a fifth card not of the same denomination as either of the pairs)

7. A committee of 6 people is to be chosen from among the Student Council, which has 12 members, 3 of whom are the President, the Secretary, and the Treasurer. Exactly two of these officers must be on the committee. In how many ways can the committee be chosen?

8. Suppose 6 factors affect the growth of a plant. Each of the factors, abundant sunlight, adequate water, and soil nutrients, causes a 1 in. increase in the plant's height after a specified time, while each of the factors, insect damage, wind, and animal browsing, causes a 1 in. decrease in this height. For any given plant, assume each of these factors is either present or absent.

    a. Suppose the normal height of the plant after the specified time is 5 in. Find the number of ways, due to the presence or absence of each of the 6 positive or negative growth factors listed, that the plant can actually reach a height of 5 in.

    b. Find all the other possible heights of the plant after the specified time, under the effects of 0 or more of the growth factors, and the number of ways the plant can attain each height.

80

Challenge Problems, ALGEBRA 2: EXPLORATIONS AND APPLICATIONS

Challenge Set 70 ···

1. **Writing** Suppose you know a particular term of a binomial expansion $ka^{n-r}b^r$. Explain why you can find the next coefficient using the following method. Multiply the coefficient k by the exponent of a, $n - r$, and divide it by one more than the exponent of b, $r + 1$.

Find the coefficient of the term containing the given powers of a and b in the expansion of $(a + b)^n$.

2. a^5b^2 3. a^3b^2 4. a^2b^4 5. a^3b^5

★ 6. **a.** Expand the product $(a + b)^5 = (a + b)(a + b)(a + b)(a + b)(a + b)$ by the following method. For each r ($r = 0, 1, 2, 3, 4, 5$), find the number of ways you can choose r "copies" of a and $5 - r$ "copies" of b from the 5 factors in parentheses, with one "copy" coming from each factor. Multiply $a^{5-r}b^r$ by this number. Add all the terms you get.

   b. Explain why this method gives the correct answer for $(a + b)^5$.

   c. Explain why the coefficients of the terms are the numbers $_5C_r$.

7. Suppose you follow a path in Pascal's triangle like the ones shown in the diagram at the right. Starting at a number along the left-hand edge of the triangle, go downward along a diagonal any number of rows. Finally turn left one number and stop.

   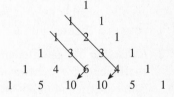

   a. State a simple arithmetic relationship among the numbers along the paths shown.

   b. State the relationship in part (a) for the two paths shown, using the $_nC_r$ notation and sigma notation. State this relationship in general. (*Hint*: Let p be the row number where the path starts, and suppose r is the number of the entry in the row where it turns left.)

★ 8. In Pascal's triangle, the entry $_7C_3$ is the sum of the entries $_6C_2$ and $_6C_3$.

   a. Use this example to write a general formula, in terms of n and r, for finding a particular entry in Pascal's triangle from the sum of two entries in the row above it.

   b. Prove the formula you wrote in part (a) by combining the two terms of the sum over a common denominator. (*Hint*: The least common multiple of $k!$ and $(k + 1)!$ is $(k + 1)!$.)

Chapter 12 Challenge Set ···································

Solve each equation for *n*. (*Sections 12.3 and 12.4*)

1. $_nC_2 = 55$

2. $_nC_4 = {}_nC_6$

3. $nC_3 = 4(_{n-2}C_2)$

4. The diagram at the right shows a circular table with 8 lettered seats equally spaced around it.

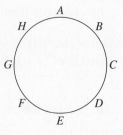

  a. In how many ways can 8 people be seated around the table?

  b. Suppose two seating arrangements are considered the same if the seats can be rotated around the table so that the two arrangements of people coincide. How many arrangements are equivalent to any particular seating arrangement by rotation? How many different seating arrangements are there, if these arrangements are counted as one?

  c. Suppose *n* people are seated around a circular table and seating arrangements that are rotations of each other are regarded as being the same. How many different arrangements are possible? (*Section 12.3*)

5. *Writing* Explain how you know that if *n* is a prime number, all the entries in the *n*th row of Pascal's triangle, except the first and the last entries, are evenly divisible by *n*. (Count the top "1" as the "0th" row.) Why doesn't your explanation hold for nonprime numbers *n*? (*Section 12.5*)

6. *Extension* The coefficients of $(a + b)^n$ can be generated as follows. The coefficient of a^n is always 1, the coefficient of $a^{n-1}b$ is always *n*, and if $k(r)$ is the coefficient of $a^{n-r}b^r$ for $r \geq 1$, then the next coefficient is

$$k(r + 1) = \frac{k(r) \cdot (n - r)}{r + 1},$$

where $n - r$ is the exponent of *a*, and $r + 1$ is 1 plus the exponent of *b*. This formula has the advantage that it makes sense for *fractional* values of *n*, although it does not produce a finite sum, so that it can be used, for example, to approximate $(1 + x)^{1/2} = \sqrt{1 + x}$.

  a. Write out the first 4 terms of $(1 + x)^{1/2}$, expanded by this method.

  b. Use your answer to part (a) to approximate $\sqrt{1.25}$. Check your result against the value you get from a calculator.

  c. Write out the first 4 terms of $(8 + x)^{1/3}$, expanded by this method. Use your result to approximate $\sqrt[3]{9}$.

Challenge Set 71

1. An archery target has rings of radius r, $2r$, and $3r$. Suppose that an arrow hits the target randomly. What is the probability that an arrow that hits the target will land in each ring?

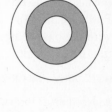

2. In a cereal factory, two machines fill boxes with two different cereals. Suppose that, as each box goes by on the assembly line, each machine deposits in it a random amount of cereal between 4 oz and 6 oz.

  a. Represent as a region in a plane all the possibilities for the amounts deposited in a box by the two machines.

  b. Shade the part of the region in part (a) that represents the possibilities that the total amount of cereal in a box will be between 9 oz and 11 oz.

  c. Find the probability that the amount of cereal in a box will be between 9 oz and 11 oz.

3. a. Find the probability of getting exactly two 6's when three dice are rolled. (*Hint*: Assume the dice are rolled one after the other.)

  b. Find the probability of getting exactly two dice that show the same number of spots.

4. A stoplight has a cycle during which it is red for 20 s and green for 12 s.

  a. Graph the amount of time you would have to wait at the light against the time during the cycle. Let 0 correspond to the time when the light turns red.

  b. Explain why the *average* time you will have to wait at the light is the *area* under the graph in part (a) divided by the duration of the cycle. Find this average time.

5. a. Write a formula that can be used to generate a random number between 0 and 2 on your calculator.

  b. Suppose you generate many random numbers between 0 and 2. Describe a way to use your results to approximate $\sqrt{2}$ experimentally, without using the square-root key.

★ **6.** The diagram at the right shows a semicircle with center at D and another semicircle with center at F, the midpoint of $\overline{BC}$. Suppose $DB = 2$.

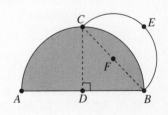

  a. Find the area of the *unshaded* region and the area of the small semicircle *BEC*.

  b. Explain how you could approximate π using an experiment in which darts are dropped onto semicircle *BEC*.

Challenge Set 72 ···

1. In her softball league, Maria Martinez has an on-base average of .400. This means that the probability is 0.4 that she will get on base in a given at-bat. Po Lan Chang has an on-base average of .450 with women on base, and .300 with the bases empty. Suppose Maria is the first batter one inning and Po Lan is the second batter. Find each probability.

   a. Exactly one of the two women gets on base.

   b. Neither of the women gets on base.

   c. At least one of the women gets on base.

2. One computer programming technique for storing data is called *hashing*. In this technique, each of certain designated memory slots can store one item of data. The computer randomly chooses one of these slots for the first data item and then randomly chooses another slot, or the same slot, for the second data item, and so on. If the same slot is chosen for two data items, a *collision* occurs, and the computer must put the second item somewhere else. Suppose there are 100 slots and 10 data items.

   a. What is the probability that no collisions occur when all 10 items are hashed?

   b. For the same 100 slots, how many data items must be hashed before the probability of at least one collision becomes greater than $\frac{1}{2}$?

3. A change purse contains 5 quarters and 6 dimes. Suppose you reach in and pull out two coins at random. Find each probability.

   a. Both coins are quarters.   **b.** Both coins are dimes.

   c. Both coins are the same denomination.   **d.** The coins are different denominations.

4. **a.** Suppose two cards are picked at random from a standard deck. Is the probability greater that the cards will be of the same color (both red or both black) or that they will be of opposite colors? Explain.

   b. Suppose that in a town with a large population, represented by $2n$, half the voters are Democrats and half Republicans. Suppose also that two voters are chosen at random. Is the probability greater that they will be of the same party or of opposite parties? Explain.

★ 5. In a certain town, 40% of the prospective buyers of used cars first try Lowball Bob's, while the other 60% first try Rusty's Auto Mall. Of those who first try Bob's, 50% also check out Rusty's; of those who first try Rusty's, 30% also check out Bob's. Suppose that every buyer who tries only one dealer buys from that dealer, and that 75% of those who check out both dealers buy from Bob. What percent of the prospective buyers end up buying their cars at Lowball Bob's?

Challenge Set 73 ···

1. a. Use the definition of conditional probability and the definition of independent events to show that two events A and B are independent if and only if $P(B) = P(B \mid A)$.

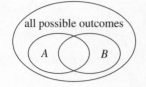

all possible outcomes

b. Use the diagram at the right to show that, for any two events A and B, $P(B) = P(B \mid A)$ if and only if $P(A) = P(A \mid B)$. (*Hint:* Let a = the number of outcomes in event A, let b = the number of outcomes in event B, let c = the number of outcomes in the overlap (A and B), and let d = the number of possible outcomes. The ratio of the area of the overlap to the area of A is then $P(B \mid A) = \dfrac{c}{a}$.)

2. Bag A contains 10 blue marbles. Bag B contains 6 blue marbles and 8 red marbles. One of the two bags is chosen at random and one marble is drawn, also randomly.

a. What is the probability of drawing a blue marble?

b. Suppose the marble drawn is blue. What is the probability that it came from Bag A?

3. A jar contains 3 red marbles and 2 green marbles. A second jar contains 4 red marbles and 8 green marbles. One marble is picked at random from each jar. Find each probability.

a. The 1st is red and the 2nd is green.   **b.** The 2nd is green, given that the 1st is red.

c. The 2nd is green.   **d.** Exactly one is green, and one is red.

e. The 1st is red, given that the 2nd is red.

4. Suppose two arrows randomly hit the archery target shown, which has rings of radius 1 ft, 2 ft, and 3 ft. Find each probability.

a. The total score is 80.   **b.** At least one arrow scores 30.

c. The second arrow scores 50, given that the first scored 10.

d. The two arrows score at least 40.

★ **5.** A stick 1 foot long is broken into two pieces.

a. What is the probability that the smaller piece has length less than $\dfrac{1}{4}$ ft?

b. Suppose, after the first break, the larger piece is broken again. Given that the first piece has length exactly $\dfrac{1}{4}$ ft, what is the probability that the three pieces can be put together to form a triangle? (*Hint:* Recall the Triangle Inequality.)

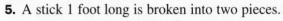

Challenge Problems, ALGEBRA 2: EXPLORATIONS AND APPLICATIONS

Challenge Set 74 ....

1. Suppose you toss a coin in groups of 3 tosses many times and you find that
2 out of the 3 tosses come up heads about $\frac{4}{9}$ of the time.

   a. If the coin were a fair coin, what should be the probability of 2 out of
3 heads?

   b. According to the results of your experiments, what seems to be the
probability p that the coin will come up heads on any given toss?

2. When a set in a professional tennis match is tied at 6 games for each player,
a tie-breaker game is played. The first player to win at least 15 points *and be
at least 2 points ahead of his or her opponent* is the winner. Suppose players
A and B are tied at 15 points each in a tie-breaker, with player A having a
60% chance of winning any given point.

   a. Use a tree diagram to find the probabilities that, after 2 more points have
been played, A will win, B will win, or the game will still be tied.

   b. Find a formula, in terms of n, for the probability that, after $2n$ more points
have been played beyond the 15–15 tie, the game will still be tied.

   c. How many points will have to be played after the 15–15 tie to make the
probability that one of the two players will win greater than 99%?

3. You are taking a 10-question true-false test. Suppose you have studied
enough so that you have a probability p of getting each question right.

   a. Write a polynomial function in the variable p for the probability that you
will score 80% or better on the test.

   b. What does p have to be in order that the probability of scoring 80% or

      better on the test is above $\frac{1}{2}$? (*Hint*: Draw a graph.)

★ **4. a.** Suppose p is a number such that $0 < p < 1$. Show that

$$\sum_{k=0}^{3} {}_3C_k \, p^k(1-p)^{3-k} = 1.$$

   b. Suppose p represents the probability of success in each of 3 binomial
trials. Interpret the result of part (a) in terms of probabilities regarding
such a binomial experiment.

   c. Generalize the formula in part (a) to a binomial experiment with n trials.

5. Suppose 4 fair coins are tossed. Find the probability that 3 or more coins
come up heads. Suppose 8 fair coins are tossed. Find the probability that 6
or more coins come up heads. Make a general statement relating n to the
probability that $3n$ out of $4n$ fair coins will come up heads when $4n$ fair
coins are tossed.

Challenge Set 75 ···

1. The weights of tomatoes picked at a farm are normally distributed with a mean of 4 oz. Suppose that in one day's harvest 29 tomatoes out of 250 weigh less than 3.4 oz. What is the standard deviation of the data?

2. In a chemistry class at a large university, grades are normally distributed. There are 152 students in the class, and the mean grade is 82. Suppose 44 students have grades between 82 and 88.

   a. Find the standard deviation of the data.

   b. In a physics course there are 75 students, the mean grade is 80, and 17 students have grades between 80 and 86. Find the standard deviation of the data.

   c. Suppose that a student's grades in physics and chemistry are independent. Find the probability that a student who is in both classes has a grade of at least 86 in chemistry and a grade of at least 82 in physics.

★ **3. a.** A fair coin is tossed 10 times. Find how many ways each number of heads between 0 and 10 can come up, and draw a histogram of these values.

   b. Suppose a fair coin is tossed 2^{10} times (the sum of the heights of the bars in the histogram in part (a)), and each number of heads comes up with frequency equal to the height of the corresponding bar in the histogram. Find the theoretical standard deviation of the data.

   c. What percent of the data items fall within 1 standard deviation of the mean? How does your answer compare with the theoretical value for a normal distribution? How could you have predicted this result from the shape of the histogram?

4. Suppose $f(x)$ is a normal curve with mean 0 and standard deviation σ. Each function given below is also a normal curve. Find the mean and the standard deviation of each function as a number or in terms of σ.

   a. $f(x + 3)$ **b.** $f(5x)$ **c.** $f(2x - 7)$

5. Suppose 20% of a set of normally distributed data, with standard deviation σ, falls between $n\sigma$ and $(n + 1)\sigma$. Find all possible values of n.

★ **6.** *Writing* The formula for the standard normal curve is $y = \dfrac{1}{\sqrt{2\pi}}e^{-x^2/2}$. Why do you think the factor $\dfrac{1}{\sqrt{2\pi}}$ is there? How would the graph be different without it?

Chapter 13 Challenge Set ·········

1. a. Suppose two people are chosen at random from a high school class. What is the probability that their birthdays do not fall on the same day of the same month? Assume that there are 365 days in a year.

   b. Suppose the birthdays of the two people already chosen are different, and a third person is chosen from the class. What is the probability that this person's birthday is different from the birthdays of the other two?

   c. How many people must be chosen before the probability is greater than $\frac{1}{4}$ that two of them have the *same* birthday? (*Section 13.2*)

2. A jar contains 3 blue marbles and 2 red marbles. A second jar contains 1 blue marble and 2 red marbles.

   a. Suppose one of the two jars is chosen at random and then a marble is selected at random from that jar. What is the probability that the selected marble is red?

   b. Suppose that, after a marble is selected from one of the two jars as in part (a), it is *returned* to the other of the two jars. Then the process of part (a) is repeated. What is the probability that the second marble is red, given that the first was red?

   c. What is the probability that both marbles are red? Do you think that the two events "the first marble is red" and "the second marble is red" are independent? (*Section 13.3*)

★ **3.** In a point in tennis, you have 2 serves to put the ball in play; if both serves go out, you lose the point. Suppose there is a 40% probability that Kichi's first serve will go in and an 80% probability that her second serve will go in, and suppose there is a 75% probability of her winning on a first serve and a 50% probability of her winning on a second serve. Find the probability that Kichi will win any given point when she is serving. (*Section 13.4*)

4. a. Write a formula that will produce one of the integers 1 or −1 on your calculator.

   b. Suppose you run the formula that you wrote in part (a) on your calculator four times and add up the results. What is the range of possible sums you could get? Find the probability of getting each of these sums. (*Section 13.4*)

5. *Extension* By keeping a record of the times she arrived at work, Lena has produced the probability distribution shown at the right, in which x represents number of minutes after 8:55. As in a normal distribution, the probability of her arriving before a time x is given by the *area* under the graph between −10 (representing 8:45) and x. What is the probability of her arriving before 9:00 ($x = 5$)? Write a piecewise defined function of x that gives the probability of her arriving before time x for $-10 \leq x \leq 0$ and for $0 \leq x \leq 10$. Make up a table of values like the table for the standard normal distribution. (*Hint*: Use similar triangles.)

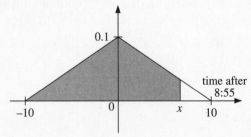

Challenge Set 76 ···

FOR USE WITH SECTION 14.1

1. **a.** Choose several acute angles A at random, and calculate the quantity $(\sin A)^2 + (\cos A)^2$. (*Note*: We usually write "$\sin^2 A$" and "$\cos^2 A$" for $(\sin A)^2$ and $(\cos A)^2$.) What do you notice? Make a conjecture about all acute angles, based on your results.

   b. Express $\sin A$ and $\cos A$ in terms of the sides a, b, and c of right triangle ABC and prove your conjecture from part (a).

2. Find $\sin 37°$ and $\cos 53°$. Find $\sin 69°$ and $\cos 21°$. What do you notice? Make a conjecture of the form $\sin \theta = \cos (\underline{\ ?\ })$. Explain how you know your conjecture is true by referring to a diagram like the one in Exercise 1.

3. In the diagram at the right, the angles and sides of the figure are as marked, and $m\angle PQS$ is 36°.

   a. Name two non-congruent similar triangles in the diagram.

   b. Use similar triangles to find an equation involving x and no other unknown lengths. Solve this equation for x. Is there more than one possible value for x that makes sense in the diagram?

   c. Use the value(s) of x that you found in part (b) to find a radical expression for $\sin 18°$. Does your answer agree with the value of $\sin 18°$ supplied by your calculator?

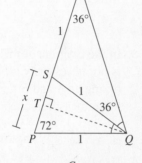

4. You can use trigonometry to approximate the number π. In the diagram at the right, $ABCDEFGHJ$ is a regular 9-sided polygon inscribed in a unit circle of radius 1 centered at P.

   a. Find the value of θ in degrees. Express a side x of the polygon in terms of θ, using sine and/or cosine. (*Hint*: Refer to $\triangle PCD$.) Express the perimeter of the polygon in terms of θ.

   b. Suppose the polygon shown had n sides, rather than 9 sides. Express θ in terms of n, and use this expression to write the perimeter of the polygon in terms of n only, using sine and/or cosine.

   c. Evaluate the expression for the perimeter that you found in part (b) for $n = 20$. The circumference of the unit circle is actually 2π. How many sides does a polygon need to have before its perimeter approximates this circumference to two decimal places?

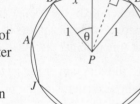

Challenge Problems, ALGEBRA 2: EXPLORATIONS AND APPLICATIONS

Challenge Set 77 ·······································

1. a. Choose several acute angles A at random and calculate the value of the expression $\dfrac{1}{\cos^2 A} - \tan^2 A$. (*Note*: Recall that "$\cos^2 A$" means $(\cos A)^2$ and similarly for "$\tan^2 A$.") What do you notice? The function $\dfrac{1}{\cos \theta}$ is called *secant* θ, denoted "sec θ." Make a conjecture about all acute angles that involves the tangent function and the secant function.

b. Prove the conjecture you made in part (a) by drawing a right triangle with angles A, B, and C and opposite sides a, b, and c.

2. When the sun is 20° above the horizon, a flagpole casts a shadow 63 ft long. How long will the pole's shadow be when the sun has sunk to an angle of 10° above the horizon?

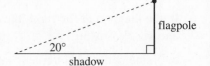

3. In the diagram at the right $\overline{CD}$ is the altitude to $\overline{AB}$ in $\triangle ABC$ which is not necessarily a right triangle. Suppose you know the measures of angles A and B and the length c. Find a formula for h in terms of these measurements. (*Hint*: Let $x = AD$. Consider the function $\dfrac{1}{\tan A} + \dfrac{1}{\tan B}$.)

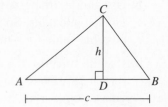

4. Use the result of Exercise 3 to solve the following problem. At a certain instant, two observers on the ground, and in the same vertical plane as a rocket, sight the rocket at angles of elevation of 70° and 80°. The distance between the observers is 0.40 mi. How high above the ground is the rocket at this instant?

5. In the diagram at the right, the segment $\overline{AB}$ is a tangent to the unit circle whose center is at the origin.

a. Explain how you know that $\overline{OP}$ is perpendicular to $\overline{AB}$.

b. Find a segment whose length is tan θ and a segment whose length is sec θ. (*Hint*: See Exercise 1.)

c. *Writing* Give a possible explanation for the choices of the names of these functions, based on your answers to part (b).

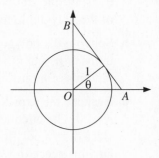

★ **6.** Suppose that θ is an acute angle and that sin $\theta = x$. Find tan θ in terms of x. (*Hint*: Draw a right triangle with θ as one of its acute angles. Label the sides with their lengths.)

Challenge Set 78 ··

The secant, cosecant, and cotangent functions are defined as follows:

$$\sec \theta = \frac{\text{hypotenuse}}{\text{adjacent}} = \frac{1}{\cos \theta}, \qquad \csc \theta = \frac{\text{hypotenuse}}{\text{opposite}} = \frac{1}{\sin \theta}, \text{ and}$$

$\cot \theta = \dfrac{\text{adjacent}}{\text{opposite}} = \dfrac{1}{\tan \theta}$. **Find the values of these three functions for an angle θ**

with the given point P on its terminal side.

1. $P(7, 24)$ **2.** $P(6, -8)$ **3.** $P(-33, -56)$ **4.** $P(-20, 21)$

Find the values of sin θ, cos θ, and tan θ for the angle θ with terminal side in the given quadrant and having the given value for one of these functions.

5. II; $\sin \theta = \dfrac{9}{41}$ **6.** IV; $\tan \theta = -\dfrac{39}{80}$ **7.** III; $\cos \theta = -\dfrac{55}{73}$

8. a. Express $\tan \theta$ and $\cot \theta$ in terms of $\sin \theta$ and $\cos \theta$.

     b. A *trigonometric identity* is an equation involving trigonometric functions that is true for every value of the variable(s) for which it makes sense. Use your answer to part (a) and the identity $\sin^2 \theta + \cos^2 \theta = 1$ to prove the following identity:

$$\tan \theta + \cot \theta = \sec \theta \csc \theta$$

9. a. For several randomly chosen values of θ, find the values of the expression $\tan^2 \theta - \sin^2 \theta$ and $\tan^2 \theta \sin^2 \theta$. What do you notice?

★   **b.** State and prove an identity based on your answer to part (a).

10. Suppose a ship starts at the origin of a coordinate system and sails a distance r to the point P shown in the diagram, then changes course and sails a distance s to Q.

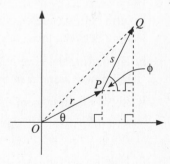

★   **a.** Find the coordinates of P and Q in terms of the distances r and s and trigonometric functions of the angles θ and ϕ.

     b. Suppose $\theta = 28°$, $\phi = 64°$, $r = 10$ mi, and $s = 12$ mi. Find the distance between the boat's final position, point Q, and its starting point, the origin, to the nearest mile.

Challenge Set 79 ···

1. a. Explain why it makes sense that the *arclength L* of a sector of a circle
with radius r and angle θ should be given by the formula

$$L = \frac{\theta}{360} \cdot 2\pi r = \frac{\theta}{180} \cdot \pi r.$$

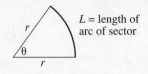

L = length of
arc of sector

b. Use the formula for the arclength of a sector of a circle from part (a) and
the formula for the area of a sector in terms of r and θ to find a formula
for the area of a sector in terms of r and arclength L only.

c. A sector of a circle has a perimeter of 14 in., including the radii, and an
area of 12 in.². Use the formula from part (b) to find all possible values of
the radius.

2. A goat is tied by a 60 m rope to the corner of a barn against a fence, as
shown.

a. Sketch the region that the goat can roam around in while remaining
tied by the rope.

b. Find the area of the region you sketched in part (a).

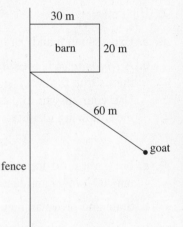

30 m

barn   20 m

60 m

•goat

3. A dog is tied by two leashes, each 8 ft long, to each of two stakes
8 ft apart.

a. Sketch the region in which the dog can roam, constrained by the
leashes.

b. Find the area of the region in part (a).

fence

★ **4.** In this exercise you will discover a formula for the area of a triangle
in terms of two angles, θ and ϕ in the diagram, and the length of the
side, c, between these angles.

a. By writing the length BD as $c - x$, set up two equations involving the
unknowns x and h in terms of c and trigonometric functions of θ and ϕ.

b. Solve the simpler of the two equations you wrote in part (a) for x and
substitute this expression for x in the other equation. Solve this equation
for h.

c. Use the formula Area $= \frac{1}{2}$(base)(altitude) to find an expression for the

area of $\triangle ABC$ in terms of c and trigonometric functions of θ and ϕ.

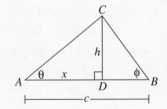

Challenge Set 80 ···

FOR USE WITH SECTION 14.5

1. a. Show that, in $\triangle ABC$, $b = \dfrac{a \sin B}{\sin A}$.

  b. Use one of the area formulas for a triangle, together with the result of part (a), to get a formula for the area of a triangle in terms of one side of the triangle and trigonometric functions of its three angles.

2. In $\triangle PQR$, $\tan P = \dfrac{3}{4}$, $\tan Q = \dfrac{15}{8}$, and $p = 6$. Find q. (*Hint*: Find $\sin P$ and $\sin Q$ first.)

3. *Writing* Suppose that in $\triangle ABC$, $\tan A = k < 0$. Explain how you know that, given the value of k and the values of a and b, there is at most one possible value for $m\angle B$.

4. Suppose that, in $\triangle ABC$, you know that $m\angle A = 23°$, and suppose that, using the law of sines to solve for $m\angle B$, you find that one solution is $m\angle B = 65°$. Explain why there must be another solution for $m\angle B$. List all the angles in both solutions of the triangle.

5. Working for a surveying team, Jackie Johnson needed to calculate the distance between two houses (A and B in the diagram) on the far shore of a lake. She measured the angles and the distance shown in the diagram, and she estimated that $ABCD$ is a trapezoid. How far apart are the two houses, to the nearest meter?

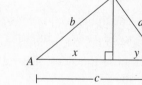

★ **6.** In this exercise, you will derive a formula for the sine of the *sum* of two acute angles, A and B. Suppose a triangle containing these angles, like the one shown, is drawn.

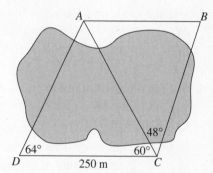

  a. Explain how you know that $\sin C = \sin (A + B)$, and use this fact to show that $\sin (A + B) = \dfrac{c}{b} \sin B$. (*Hint*: How is $\sin \theta$ related to $\sin (180 - \theta)$?)

  b. Show that $c = a \cos B + b \cos A$. Substitute this expression for c in the equation in part (a) and simplify your result.

  c. Explain how you know that $\dfrac{a}{b} = \dfrac{\sin A}{\sin B}$, and use this fact to eliminate a and b from your result in part (b). Express $\sin (A + B)$ in terms of $\sin A$, $\cos A$, $\sin B$, and $\cos B$.

Challenge Problems, ALGEBRA 2: EXPLORATIONS AND APPLICATIONS

Challenge Set 81 ....

1. **a.** Find the value of the expression $n\sqrt{2 - 2\cos\left(\dfrac{360}{n}\right)}$ for $n = 20, 30,$ and 50.

What value do these numbers seem to approach? (*Hint:* Divide each number by 2.)

b. Explain your answer to part (a) by drawing a regular polygon with n sides inscribed in a circle of radius 1 centered at the origin of a coordinate system, and considering the perimeter of the polygon.

2. You can find x in the diagram at the right, without looking up any values of the trigonometric functions or getting their values from a calculator.

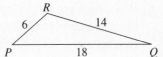

a. Find $\cos A$ as a rational number using the law of cosines applied to $\triangle ABC$.

b. Use the result of part (a) to find x, again making use of the law of cosines.

3. Use the method of Exercise 2 to find the length of the median drawn to side $\overline{PQ}$ in $\triangle PQR$, with the measurements shown. (*Hint:* Recall that a median joins a vertex of a triangle with the midpoint of the opposite side.)

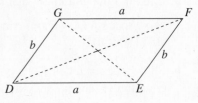

4. **a.** Find an equation relating $\cos A$ and $\cos (180 - A)$, for any acute angle A. Does your equation also hold if A is obtuse? (*Hint:* Sketch the two angles in standard position.)

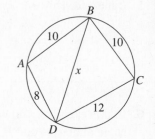

b. Use the law of cosines and the result of part (a) to find two expressions for x^2 in the diagram at the right, both in terms of $\cos A$. (*Hint:* Recall that opposite angles of a quadrilateral inscribed in a circle are supplementary.)

c. Use the two equations you found in part (b) to find $\cos A$. Substitute the value of $\cos A$ in either of your equations to find x.

5. Suppose the sides of parallelogram $DEFG$ have lengths a and b, as shown.

a. Use the law of cosines and the result of part (a) of Exercise 4 to express the squares of the lengths of the diagonals, $(DF)^2$ and $(EG)^2$, in terms of a, b, and $\cos D$.

b. Use the result of part (a) to show that the sum of the squares of the lengths of the diagonals of a parallelogram equals the sum of the squares of the lengths of its sides.

Chapter 14 Challenge Set ·······················

1. In this exercise, you will derive a formula for cos (α + β), where α and β are acute angles whose sum is less than 90°. (The formula actually works for any α and β.) In the diagram, note that $OP = 1$, and assume that ∠ORP is a right angle. (*Section 14.1*)

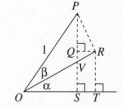

  a. Explain how you know that $OS = \cos(\alpha + \beta)$.

  b. Find an expression for OT in terms of trigonometric functions of α and β. (*Hint*: First express OR in terms of α and/or β, and then use the result to express OT in the same manner.)

  c. Explain how you know that $m\angle QPR = \alpha$. Use this fact to find an expression for ST, in terms of trigonometric functions of α and β. (*Hint*: Note that $ST = QR$. First express PR in terms of α and/or β, and then use the result to express ST likewise.)

  d. Use the fact that $OS = OT - ST$ to write a formula for cos (α + β), in terms of trigonometric functions of α and β.

2. When two forces pull on an object in two different directions, you can find the magnitude and direction of the *resultant* , or combined, force on the object using a *vector diagram*, like the one shown. In this diagram $ABCD$ is a parallelogram, $AB = 50$ lb, $AD = 40$ lb, and ∠DAB, the actual angle between the directions of the two forces, is 62°. (*Section 14.5 and 14.6*)

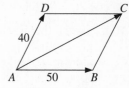

  a. Use the law of cosines to find AC, the magnitude of the resultant force.

  b. Use the law of sines to find $m\angle CAB$, the angle at which this force acts.

3. *Extension* Astronomers use a method called *triangulation* to measure the distance to an object that is relatively close to Earth. Observations are made from two points on Earth, such as A and B in the diagram. The angles α and β, the deviations from the vertical of the lines of sight, are measured, and θ, the difference between the latitudes of the two observation points if they lie on the same longitude, is determined. Assume that Earth is a sphere with center Q and radius r. Describe a procedure that could be used to find the distance AP, and use your procedure to find this distance if θ = 42°, α = 23°, β = 24°, and $r = 4000$ mi.

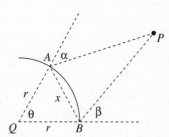

Challenge Set 82 ···

1. The sine of the sum of two angles, A and B is given by the formula
   $$\sin(A + B) = \sin A \cos B + \sin B \cos A.$$

   Use this formula to prove each of the following identities, and illustrate each with a unit-circle diagram.

   a. $\sin(90 + \theta) = \cos\theta$ **b.** $\sin(180 - \theta) = \sin\theta$ **c.** $\sin(\theta - 90) = -\cos\theta$

2. **a.** Use the formula from Exercise 1 to write $\sin(\theta - 45)$ in terms of trigonometric functions of θ.

   b. *Writing* Graph the function you wrote in part (a) and describe the relationship of the graph to the graph of $y = \sin\theta$ as well as its relationship to $y = \cos\theta$.

Find all values of θ, for $0° \le \theta < 360°$, that satisfy each equation. Give answers to the nearest degree

3. $\sin^2\theta = 0.5$ 4. $1 - \cos^2\theta = 0.4$ 5. $4\sin^2\theta - 1 = 2$

6. $\cos 2\theta = 0.788$ (*Hint*: Note that $0° \le \theta < 360°$ when $0° \le 2\theta < 720°$.)

7. **a.** Sketch the graph of $y = \sin\theta$ for $0° \le \theta \le 360°$. Also sketch some tangent lines to the graph, and on the same axes sketch the graph of the equation $y =$ (the slope of the tangent line to the graph of $y = \sin\theta$).

   b. What familiar function does your graph of the slopes resemble? What do you notice about the graph of the slopes at a point where the graph of $y = \sin\theta$ has a local maximum or a local minimum?

★ 8. In a car engine, explosions in the cylinder push the pistons down, and this force is transmitted to the crankshaft, which drives the wheels, by a connecting rod. Suppose the radius of the crankshaft, OA, is 3 in. and the length of the connecting rod, AP, is 8 in.

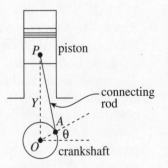

   a. Use the law of cosines to write an equation relating the distance y between the bottom of the piston P and the center of the crankshaft O, in terms of $\sin\theta$. (*Hint*: Note that $\angle AOP = 90 - \theta$.)

   b. Use the quadratic formula to solve the equation you found in part (a) for y as a function of θ. Graph this function.

Challenge Set 83 ···

FOR USE WITH SECTION 15.2

1. Suppose you are given a number y between -1 and 1. Explain, using a unit-circle diagram, why there is exactly one radian measure θ between $-\dfrac{\pi}{2}$ and $\dfrac{\pi}{2}$ such that $\sin \theta = x$. Give a range of θ that produces the same result for the cosine function, and illustrate why this is so with a unit-circle diagram.

2. The function $\sin^{-1} x$ assigns to a number x the radian measure θ, from the range in part (a), such that $\sin \theta = x$. The function $\cos^{-1} x$ is defined similarly. Find each value.

   a. $\sin^{-1}(-0.5)$ **b.** $\cos^{-1} 0.7071$ **c.** $\cos^{-1}\left(-\dfrac{\sqrt{2}}{2}\right)$ **d.** $\sin^{-1}(-1)$

Solve each equation for θ in radians, where $0 \le \theta < 2\pi$.

3. $2\cos^2 \theta + \cos \theta - 1 = 0$ **4.** $4\sin^2 \theta - 3 = 0$

5. $2\sin^2 \theta + \sin \theta = 0$ **6.** $\cos^2 \theta - \sin^2 \theta = 0$

                              (Hint: Use the identity: $\sin^2 \theta + \cos^2 \theta = 1$.)

7. a. Using a calculator set to radian measure, compute the values of the function $\dfrac{\sin \theta}{\theta}$ for the following radian measures θ: $0.3, 0.2, 0.1, 0.05$.

   b. Make a conjecture based on your answers to part (a).

8. a. Let $f(x) = x - \dfrac{x^3}{6}$. Compare the values of $f(x)$ and $\sin x$ for the following values of x: $0.1, 0.2, 0.5, 0.7$. What do you notice?

   b. Graph the functions in part (a) on the same axes. Based on your graphs, find a value of x for which your observation of part (a) does not hold.

   c. Let $g(x) = f(x) + \dfrac{x^5}{120} = x - \dfrac{x^3}{6} + \dfrac{x^5}{120}$. Add the graph of $g(x)$ to the two graphs you drew in part (b). How does the relationship between $g(x)$ and $\sin x$ compare with the relationship between $f(x)$ and $\sin x$?

9. *Writing* Graph the functions $g(x) = 1 - \dfrac{x^2}{2} + \dfrac{x^4}{24}$ and $\cos x$ on the same axes.

Based on your graphs, make any predictions you can about the values of these two functions.

Challenge Set 84 ·····································

1. a. The sine and cosine of the sum of two angles α and β are given by the formulas:

$$\sin(\alpha + \beta) = \sin\alpha\cos\beta + \sin\beta\cos\alpha \qquad \cos(\alpha + \beta) = \cos\alpha\cos\beta - \sin\alpha\sin\beta$$

Use these formulas to derive formulas for $\sin 2\theta$ and $\cos 2\theta$ in terms of trigonometric functions of θ.

b. Use the identity $\sin^2\theta + \cos^2\theta = 1$ to rewrite the formula for $\cos 2\theta$ you found in part (a) in two other ways, (1) as a function of $\sin\theta$ alone and (2) as a function of $\cos\theta$ alone.

2. Let γ equal 2θ, so that $\theta = \frac{\gamma}{2}$. Substitute $\frac{\gamma}{2}$ for θ in the two formulas you

found in part (b) of Exercise 1 to derive formulas for $\sin\frac{\gamma}{2}$ and $\cos\frac{\gamma}{2}$ in terms

of trigonometric functions of γ. These formulas contain a "±" sign. Explain how you can tell which of the two signs applies in a particular case.

Use the formulas you found in Exercises 1 and 2 and any identities you have proved previously to prove each identity.

3. $(\sin\theta + \cos\theta)^2 = 1 + \sin 2\theta$

4. $\dfrac{1}{\cos^2\left(\frac{\theta}{2}\right)} + \dfrac{1}{\sin^2\left(\frac{\theta}{2}\right)} = \dfrac{4}{\sin^2\theta}$

5. a. The *frequency* of a wave is the reciprocal of the period of the wave, expressed in cycles per second, or Hertz (Hz). An AM radio wave consists of a high-frequency *carrier wave* whose amplitude varies according to the sound it is transmitting. For example, in the equation $y = (\sin 5\pi t)(\sin 50\pi t)$, the function $\sin 5\pi t$ is the "amplitude" of the wave, corresponding to a in the formula $y = a\sin bx$.

a. Graph the function $y = (\sin 5\pi t)(\sin 50\pi t)$, and find the frequencies of the sound wave and the carrier wave.

b. Suppose a radio station wants to transmit a sound with a frequency of 220 Hz, and its carrier wave has a frequency of 840 kHz, that is, 840,000 Hz. Write an equation of the wave as a function of time t, in seconds.

6. a. Graph the function $y = x\sin 10x$. Explain the shape of the graph in terms of amplitude.

b. Predict the shape of the graph of $y = x^2\sin 10x$. Graph this function to check your prediction.

Challenge Set 85 ··

Use the formulas for the sine and cosine of the sum of two angles,

$$\sin (\alpha + \beta) = \sin \alpha \cos \beta + \sin \beta \cos \alpha$$
$$\cos (\alpha + \beta) = \cos \alpha \cos \beta - \sin \alpha \sin \beta,$$

to write each function in terms of $\sin \theta$ and/or $\cos \theta$.

1. $y = 2 \sin \left(\theta + \dfrac{\pi}{3} \right)$ **2.** $y = -3 \cos \left(\theta + \dfrac{\pi}{4} \right)$ **3.** $y = 5 \sin \left(\theta + \dfrac{5\pi}{6} \right)$

4. Use a unit-circle diagram with an angle in Quadrant I to illustrate each statement that holds for any angle θ.

    a. $\sin (-\theta) = -\sin \theta$ **b.** $\cos (-\theta) = \cos \theta$

5. Use the results of Exercise 4 together with the formulas for $\sin (\alpha + \beta)$ and $\cos (\alpha + \beta)$ to prove each identity.

    a. $\sin (\alpha + \beta) + \sin (\alpha - \beta) = 2 \sin \alpha \cos \beta$

    b. $\cos (\alpha + \beta) + \cos (\alpha - \beta) = 2\cos \alpha \cos \beta$

6. Use the results of Exercise 5 to rewrite the function $y = \sin 3\theta \cos 2\theta$ as a *sum* of sines and/or cosines. Graph both the original function and the rewritten function in the same coordinate system. Are their graphs the same?

★ **7. a.** Suppose p and q are two nonzero real numbers such that $p^2 + q^2 = 1$. Explain how you know that there is an angle θ such that $\cos \theta = p$ and $\sin \theta = q$.

    b. Suppose a and b are any two real numbers with not both 0. Show that

        $p = \dfrac{a}{\sqrt{a^2 + b^2}}$ and $q = \dfrac{b}{\sqrt{a^2 + b^2}}$ satisfy the equation in part (a).

    c. Suppose a function of the form $y = a \sin \alpha + b \cos \alpha$ is given, and θ is an

        angle such that $\cos \theta = \dfrac{a}{\sqrt{a^2 + b^2}}$ and $\sin \theta = \dfrac{b}{\sqrt{a^2 + b^2}}$. Use the results of

        parts (a) and (b) to rewrite the function in the form $y = k \sin (\alpha + \theta)$,

        where k is a constant. (*Hint:* Start by dividing both sides by $\sqrt{a^2 + b^2}$.)

8. *Writing* Graph the functions $y = \sin \theta + \sin (\theta + k)$ for $k = \dfrac{\pi}{4}, \dfrac{\pi}{2}, \dfrac{2\pi}{3}$, and $\dfrac{5\pi}{6}$

and describe the relationship between k and the amplitude of the corresponding function.

Challenge Set 86 ·····················

Prove each identity.

1. $\tan^2 \theta - \sin^2 \theta = \sin^2 \theta \tan^2 \theta$ (*Hint*: $\tan \theta = \dfrac{\sin \theta}{\cos \theta}$.)

2. $\dfrac{\tan^2 \theta}{1 + \tan^2 \theta} = \sin^2 \theta$

3. $\tan \theta + \dfrac{1}{\tan \theta} = \dfrac{1}{\sin \theta \cos \theta}$

Solve each equation for $0 \le \theta < 2\pi$.

4. $\tan^2 \theta - \tan \theta = 0$

5. $3 \tan^2 \theta - 1 = 0$

6. a. Use the formulas $\sin 2\theta = 2 \sin \theta \cos \theta$ and $\cos 2\theta = \cos^2 \theta - \sin^2 \theta$ to show that

$$\frac{1}{\tan 2\theta} = \frac{1}{2}\left(\frac{1}{\tan \theta} - \tan \theta\right).$$

(*Hint*: $\tan 2\theta = \dfrac{\sin 2\theta}{\cos 2\theta}$.)

b. Use your result from part (a) to show that $\tan 2\theta = \dfrac{2 \tan \theta}{1 - \tan^2 \theta}$.

7. a. Use the graph of the $y = \tan \theta$ to explain how you know that for any real number x there is exactly one radian measure θ, $-\dfrac{\pi}{2} < \theta < \dfrac{\pi}{2}$, such that $\tan \theta = x$.

b. Let the function $\tan^{-1} x$ be defined by $\tan^{-1} x = \theta$, where θ is as given in part (a). Graph this function.

c. Describe the end behavior of the function $\tan^{-1} x$ using infinity notation.

d. For $-1 < x \le 1$, the function $\tan^{-1} x$ is represented by the following infinite series:

$$\tan^{-1} x = x - \frac{x^3}{3} + \frac{x^5}{5} - \frac{x^7}{7} + \ldots .$$

Use this formula to show that $\dfrac{\pi}{4} = 1 - \dfrac{1}{3} + \dfrac{1}{5} - \dfrac{1}{7} + \ldots .$

★ **8.** For several values of x, find the value of the expression $\tan^{-1} x + \tan^{-1}\left(\dfrac{1}{x}\right)$.

Make a conjecture based on your results. Use a right triangle to prove your conjecture for $x > 0$.

Chapter 15 Challenge Set ································

1. In 1811, the French mathematician
Jean Baptiste Fourier made a
conjecture that any wave function,
such as the one shown, could be
represented as an infinite sum of sine
and cosine functions (called, in his
honor, a *Fourier series*):

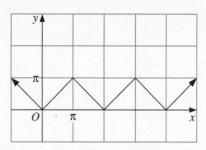

$$f(x) = a_0 + a_1 \cos x + a_2 \cos 2x + a_3 \cos 3x + \ldots + b_1 \sin x + b_2 \sin 2x + b_3 \sin 3x + \ldots$$

a. For the wave function shown, the only nonzero coefficients are $a_0 = \dfrac{\pi}{2}$ and

the *odd* coefficients a_{2n+1}, which are given by $a_{2n+1} = \dfrac{-4}{\pi(2n+1)^2}$. Write

out the sum of the first four nonzero terms of the series, and graph this sum.

b. By substituting an appropriate value for x in the series described in

part (a), write $\dfrac{\pi^2}{8}$ as an infinite sum of rational numbers. (*Section 15.3*)

2. a. A complex number $a + bi$ can be specified by giving its magnitude
(r in the diagram) and the angle made by a segment joining the
corresponding point in the complex plane to the origin (θ in the diagram).
Use the diagram to show that you can write

$$a + bi = r(\cos \theta + i \sin \theta).$$

(The right-hand side is called the *polar form* of a complex number.)
(*Section 15.4*)

b. Suppose you have two complex numbers in polar form: $z = r(\cos \alpha + i \sin \alpha)$ and
$w = s(\cos \beta + i \sin \beta)$. Calculate the product zw by ordinary complex multiplication.

c. Use the formulas for $\sin (\alpha + \beta)$ and $\cos (\alpha + \beta)$ to show that
$$zw = rs[\cos (\alpha + \beta) + i \sin (\alpha + \beta)].$$

State in general how the magnitude and the angle of a product of two
complex numbers are related to the magnitudes and angles of the two factors.

d. Use the result of part (c) to find $(1 + i\sqrt{3})^5$.
(*Hint*: $x^5 = x \bullet x \bullet x \bullet x \bullet x$.)

3. **Extension** The curve traced by a fixed point P on the
circumference of a unit circle as it rolls without slipping along
the x-axis, is called a *cycloid*, and has parametric equations
$x = t - \sin t$, $y = 1 - \cos t$. Graph this curve for $0 \le t \le 2\pi$, and
use trigonometry to derive the parametric equations. How do the
equations change if the circle has radius r, where r is not
necessarily equal to 1?

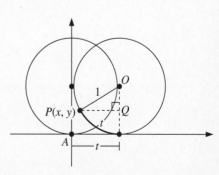

Answers ·················

Chapter 1

Challenge Set 1

1. 9.8, 25.1, 55.3, 116.1; 2.69, 2.61, 2.36, 2.21

2. Answers may vary. A sample is given. Find the range of values in each case. Divide the range of increases by the mean of the data to get a relative value; compare this to the range of growth factors.

3. –45.8; 2.35; The first figure is not likely, being negative.

4. Add one-half the average increase to the figure for 1959-1960; 41.4

5. Take the square root of the average growth factor and multiply this by the figure for 1969-1970 ; 64.0

6. 51.6

7. The yearly growth figures are $b - a$, $c - b$, $d - c$, and $e - d$. Their average is

$$\frac{(b - a) + (c - b) + (d - c) + (e - d)}{4} =$$

$$\frac{-a + (b - b) + (c - c) + (d - d) + e}{4} = \frac{e - a}{4}.$$

8. 2.8; 1.01; Add the decline as a negative number in averaging; add the growth factor as usual in averaging (the percent change for the last two years will be less than 0).

9. Use the absolute value of the increase or decrease in the average; use the absolute value of the percent increase or decrease in the average.

Challenge Set 2

1. $y = 248$; $y = 248(1.04)^x$; Answers may vary. A sample is given. I would prefer the exponential model, since it shows positive growth in sales.

2. 3.0

3. $y = 130.7(1.0375)^x$; 108.7; –5

4. $y = 4.6x + 130.7$; 107.7

5. The linear model is better. Answers may vary. A sample is given. Although the growth in CPI is not constant from year to year, the long term behavior seems to match a constant growth model fairly well.

6. **a.** \$29,000; $y = 29,000x + 155,000$; \$387,000

   b. 1.15; $y = 155,000(1.15)^x$; about \$474,000

7. $y = 4,200,000\left(\sqrt{\sqrt{\sqrt{k}}}\right)^x$

Challenge Set 3

1. $x = \dfrac{5}{2}$; $y = -1$; $z = 3$; $w = -\dfrac{7}{2}$

2. $a = 5$; $b = -6$; $c = -2$; $d = 7$

3. $\begin{bmatrix} -2 & 2 \\ -2 & -6 \\ 8 & 0 \end{bmatrix}$

4. $\begin{bmatrix} 9 & -9 \\ -6 & 12 \\ 5 & 6 \end{bmatrix}$

5. $\begin{bmatrix} 4 & -2 \\ 0 & 2 \\ -2/3 & 2 \end{bmatrix}$

6. $\begin{bmatrix} 5 & -2 \\ -1 & 0 \\ 3 & 3 \end{bmatrix}$

7. $\begin{bmatrix} 1/2 & -1/2 \\ -1/2 & 3/2 \\ -2 & 0 \end{bmatrix}$

8. $\begin{bmatrix} 3 & -17/5 \\ -6/5 & 6 \\ -7/5 & 8/5 \end{bmatrix}$

9. **a.** Since $\overline{PQ} \parallel \overline{OR}$ and $PQ = OR$, the two segments have the same rise and run.

   b. $(a + c, b + d)$

c. Area of triangle $OYQ = \frac{1}{2}(a + c)(b + d) = \frac{1}{2}(ab + ad + bc + cd)$; Area of triangle $OXP = \frac{1}{2}(ab)$; Area of triangle $PWQ = \frac{1}{2}(cd)$; Area of rectangle $XYWP = bc$; therefore, the area of triangle $OPQ = \frac{1}{2}(ad + bc) - bc = \frac{1}{2}(ad - bc)$, and area of parallelogram $OPQR = ad - bc$.

10. The entries of matrix Y are the negatives of those of matrix X, respectively.

Challenge Set 4

1. $\begin{bmatrix} -10 & 6 \\ 6 & -3 \end{bmatrix}, \begin{bmatrix} -10 & 4 \\ 9 & -3 \end{bmatrix}, \begin{bmatrix} -20 & 8 \\ 12 & -4 \end{bmatrix}, \begin{bmatrix} -20 & 8 \\ 12 & -4 \end{bmatrix},$

$\begin{bmatrix} -15 & 22 \\ 9 & -12 \end{bmatrix}, \begin{bmatrix} -21 & 8 \\ 18 & -6 \end{bmatrix}; AC = CA$

2. $\begin{bmatrix} 8 & 0 \\ 0 & 12 \end{bmatrix}, \begin{bmatrix} 8 & 0 \\ 0 & 12 \end{bmatrix}, \begin{bmatrix} 6 & -4 \\ 0 & 18 \end{bmatrix}, \begin{bmatrix} 6 & -6 \\ 0 & 18 \end{bmatrix}; BC = CB$

3. $\begin{bmatrix} 12 & -8 \\ 0 & 24 \end{bmatrix}, \begin{bmatrix} 12 & -8 \\ 0 & 24 \end{bmatrix}$; yes

4. It is sufficient that the matrix be a scalar multiple of the identity;

$\begin{bmatrix} k & 0 \\ 0 & k \end{bmatrix} \bullet \begin{bmatrix} a & b \\ c & d \end{bmatrix} = \begin{bmatrix} ka & kb \\ kc & kd \end{bmatrix} = \begin{bmatrix} a & b \\ c & d \end{bmatrix} \bullet \begin{bmatrix} k & 0 \\ 0 & k \end{bmatrix}$

5. a. $\begin{bmatrix} ap + br & aq + bs \\ cp + dr & cq + ds \end{bmatrix}$; $apcq + apds + brcq + brds - (cpaq + cpbs + draq + drbs) = apds + brcq - cpbs - draq$

b. $ps - rq$; $(ad - bc)(ps - rq) = adps - adrq - bcps + bcrq$; yes

6. a. $\begin{bmatrix} 7 & -3 \\ 5 & -2 \end{bmatrix}$

b. $\begin{bmatrix} -3/2 & -5/2 \\ -1 & -2 \end{bmatrix}$

c. $\begin{bmatrix} -3/2 & 1 \\ 1 & -1/2 \end{bmatrix}$

7. Answers may vary; for example:

$\begin{bmatrix} 1 & 0 \\ 0 & 0 \end{bmatrix}, \begin{bmatrix} 1 & 1 \\ 0 & 0 \end{bmatrix}$.

8. Multiply the entries along diagonals from upper left to lower right, wrapping around when necessary. Add these terms. Multiply entries from lower left to upper right, and subtract these terms from the sum.

Challenge Set 5

1. a. Choose a random number between 0 and 1 on a calculator. A number less than or equal to 0.6 means the player has made the free throw. If this happens, choose another random number between 0 and 1.

b. Answers may vary. A sample is given. P(no free throws) = 0.4, P(1 free throw) = 0.4, P(2 free throws) = 0.2

2. Have a calculator choose 2 random numbers between 0 and 1, say a and b. Assume $b \geq a$. The lengths of the pieces are then a, $b - a$, and $1 - b$. Then the three pieces form a triangle if $a + (b - a) > 1 - b$ or $b > 1 - b$, $a + (1 - b) > b - a$ or $b < a + \frac{1}{2}$; and $(b - a) + (1 - b) > a$ or $1 - a > a$.

3. a. Have a calculator choose 5 random numbers between 0 and 1. Choose the smallest of the 5 numbers and multiply it by 3. This is the time you will have to wait in that instance.

b. Answers may vary. A sample is given. Time in seconds: 39.0, 16.5, 11.6, 15.3, 1.29, 14.4, 31.0, 4.56, 2.05, 79.6; Average length of time in seconds: 21.5

4. a. Flip a coin, letting heads represent winning the volley and tails represent losing. Flip until you get tails. Record the number of flips in each sequence.

b. Answers will vary; should be about 2.

5. a. Have a calculator choose 3 random numbers between 0 and 1. The integer in each case will be the integer part of $10x$, where x is the number on the calculator.

b. Answers may vary; the probability should be about $\frac{1}{6}$.

c. Use the same procedure but do not take the integer part. The probability should still be about $\frac{1}{6}$.

Chapter 1 Challenge Set

1. a. 512, 538, 564, 593, 622

  b. $y = 512(1.05)^x$; very good

2. a.

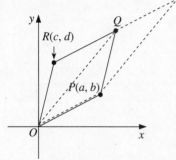

The point R falls on the old point Q. The area of the new parallelogram equals that of the original parallelogram.

  b. $ad - bc$; they are equal; yes

3. $\begin{bmatrix} a & b & c \\ 0 & d & e \\ 0 & 0 & f \end{bmatrix} \begin{bmatrix} k & m & n \\ 0 & p & q \\ 0 & 0 & r \end{bmatrix} =$

$\begin{bmatrix} ak & am + bp & an + bq + cr \\ 0 & dp & dq + er \\ 0 & 0 & fr \end{bmatrix}$

4. $\begin{bmatrix} 1/a & 0 & 0 \\ 0 & 1/b & 0 \\ 0 & 0 & 1/c \end{bmatrix}$; both products equal the identity matrix.

5. Answers may vary. A sample is given. Throw a die to determine the winner of each set, with 1, 2, 3, or 4 representing a win for Delgado. The simulation should show that the probability of Delgado winning a match is greater than $\frac{2}{3}$. (The theoretical value is about 0.79.)

Chapter 2

Challenge Set 6

1. 37.5

2. 210

3. 72

4. 195

5. $6ab$

6. $2b$

7. a. 250,000

  b. 42,000 km

8. 61.1 in.^{2}

9. a. Since (kx, ky) is on the graph, choose any point (x, y) and let $k = 0$.

  b. Since $(0, 0)$ is on the graph, $0 = m \cdot 0 + b$. Therefore, $0 = b$.

  c. Since $\left(\frac{2}{3}, 5\right)$ is on the graph,

  $\left(6 \cdot \frac{2}{3}, 6 \cdot 5\right) = (4, 30)$ is also on the graph

  ($k = 6$). Since a direct variation is a function, there is no other point on the graph with $x = 4$.

Challenge Set 7

1. $\frac{3}{5}$

2. $-\frac{5}{4}$

3. $-\frac{1}{9}$

4. $\frac{1}{3}$

5. $y = -\frac{2}{3}x + \frac{7}{3}$

6. $y = \frac{2}{5}x + 6$

7. $y = \frac{a}{4}x - 4$

8. $y = -\frac{3}{2b}x + \frac{1}{2}$

9. $y = -\frac{3}{2}x + 7$

10. The equation is unchanged.

11. a. $\frac{1}{2}$ h; 20

  b. $y = \frac{4}{3}x - \frac{100}{3}$

12. a. $(-2, 1)$, $(6, 7)$

  b. $y = \frac{3}{4}x + \frac{5}{2}$

13. $\frac{y-d}{x-c} = k$ is equivalent to $y = kx + (d - kc)$; the slope is k, and the y-intercept is $d - kc$.

Challenge Set 8

1. $\frac{7}{2}$

2. $\frac{23}{4}$

3. $\frac{47}{8}$

4. $\frac{24}{5}$

5. $-\frac{a+3}{4} + 5$

6. $-\frac{1}{2}c + 5$

7. $-\frac{1}{12}b + 5$

8. $-\frac{k}{4} + \frac{41}{8}$

9. $f(g(x)) = 2x + 3$; $g(f(x)) = 2x + 1$; no

10. $f(g(x)) = x + 2$; $g(f(x)) = x + 2$; yes

11. $f(g(x)) = -\frac{3}{4}x + 6$; $g(f(x)) = -\frac{3}{4}x - 11$; no

12. $f(g(x)) = -4x + 1$; $g(f(x)) = -4x + 6$; no

13. They have the same slope; this slope is the product of the slopes of f and g.

14. domain: all real numbers; range: $y \geq 0$

15. domain: $x \geq 2$; range: $y \leq 0$

16. domain: all real numbers; range: $y \leq 3$

17. 1, 5

18. $\frac{3}{2}$, $-\frac{19}{2}$

19. 13, -7

20. $x < -5$ or $x > 3$

21. $2 \leq x \leq 8$

22. $-2 < x < 5$

23. a^2, b^2, $a^2 + 2ab + b^2$

24. $2a + 5$, $2b + 5$, $2a + 2b + 5$

25. $3a$, $3b$, $3a + 3b$

26. $\frac{1}{a}$, $\frac{1}{b}$, $\frac{1}{a+b}$

27. True only for Ex. 25. In a direct variation $f(a + b) = f(a) + f(b)$.

Challenge Set 9

1. a–b.

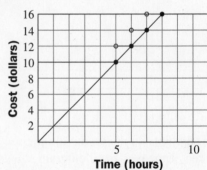

  b. The line $y = 2x$ is below the step function graph. The charges are less using the formula $y = 2x$ for all amounts of time, except for integral amounts of hours of 5 or greater, when the charges are equal.

2. $y = 3.25x + 6.5$

3. a–b.

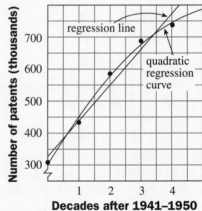

Yes; the quadratic regression curve is a better fit.

4. $y = 308(1.25)^x$; $y = 336(1.25)^x$;

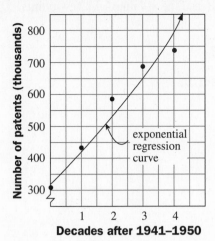

Number of patents (thousands) vs Decades after 1941–1950

exponential regression curve

Both the linear model and the quadratic model give a better fit.

Challenge Set 10

1. negative, positive, positive, positive

2. a. The products are mostly positive for a strong positive correlation and mostly negative for a strong negative correlation. The sum, and hence the mean, of the these products follows the same pattern.

b. $\left(\dfrac{x_1 - (x_1 + x_2)}{2}\right)\left(\dfrac{y_1 - (y_1 + y_2)}{2}\right) +$

$\left(\dfrac{x_2 - (x_1 + x_2)}{2}\right)\left(\dfrac{y_2 - (y_1 + y_2)}{2}\right) = x_1 y_1 + x_2 y_2 -$

$\dfrac{x_1 + x_2}{2}(y_1 + y_2) - (x_1 + x_2)\dfrac{y_1 + y_2}{2} +$

$\dfrac{2(x_1 + x_2)}{2} \cdot \dfrac{y_1 + y_2}{2} = (x_1 y_1 + x_2 y_2) - (x_1 + x_2) \cdot$

$\dfrac{y_1 + y_2}{2}$; therefore, the mean of these equals

$\dfrac{x_1 y_1 + x_2 y_2}{2} - \dfrac{x_1 + x_2}{2} \cdot \dfrac{y_1 + y_2}{2} = \overline{xy} - \overline{x} \cdot \overline{y}.$

3. 0.93

Challenge Set 11

1. 10 ft/s

2. $\dfrac{5}{2}$ ft/s

3. $\dfrac{25}{6}$ ft/s

4. $\dfrac{29}{5}$ ft/s

5.

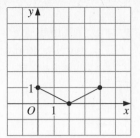

6.

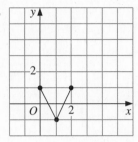

7.

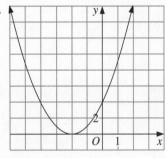

8. $y = -5t + 1$

9. $x = \dfrac{1}{4}t - \dfrac{7}{4}$

10. $y = -18t + 14$

11. $y = -\dfrac{1}{2}x + 5$; $y = -2x + 10$;

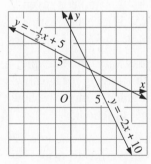

The slopes are reciprocals; the x-intercept of one graph is the y-intercept of the other, and vice versa.

12. a. $x = 40t$; $y = 70 - 30t$

b. 65 mi

1. a.

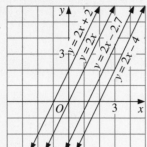

All lines are parallel to the line $y = 2x$.

b.

All lines pass through $(-2, 3)$.

c.

All lines are parallel to the line $y = -\frac{2}{3}x$, or all are horizontal translations of $y = -\frac{2}{3}x + 5$.

2. a. 8; 14

b. $y = 1.73x + 0.13$; yes

c. Answers may vary. A sample is given.
(1, 24), (2, 30), (5, 50), (7, 60), (10, 82);
$(\bar{x}, \bar{y}) = (5, 49.2)$; $y = 6.37x + 17.35$; yes

3. a. weak; answers may vary; for example, x might be hours of studying for a test; y might be the test grade; z might be how long it takes to get to school.

b. Yes; answers may vary; for example, reverse y and z in the example in part (a).

4. $\dfrac{h(t_0) - q}{g(t_0) - p}$

5. Answers may vary. A sample is given. In general, the more data points used, the more closely the regression line should match the original line. To make the slopes match more closely, move one or more points closer to the line. The y-intercept value should decrease if you decrease the y-values of a few points.

Chapter 3

Challenge Set 12

1. 2

2. 5

3. 4

4. 3

5. 5

6. 2

7. a.

Stage	Length	Width	Height	Volume
0	12	12	12	1728
1	12	6	12	864
2	6	6	12	432
3	6	6	6	216
4	6	3	6	108
5	3	3	6	54

b. $y = 1728\left(\dfrac{1}{2}\right)^n$

c. $y = a^3\left(\dfrac{1}{2}\right)^n$

8. a. 2^n

b. $y = \left(\dfrac{2}{3}\right)^n$

Challenge Set 13

1. 32

2. 5

3. 3

4. 25

5. 8

6. 100

7. 4

8. $\dfrac{1}{625}$

9. $\dfrac{1}{27}$

10. -6

11. $\dfrac{7}{2}$

12. -5

13. a. $1.05; \dfrac{t}{4}$

   b. $y = 135,000(1.05)^{t/4}$

   c. about 140,000

14. $(-4)^{3/5}$ must equal $[(-4)^3]^{1/5} = -64^{1/5}$. There is a number b such that $b^5 = -64$, but b must be less than 0. If $\dfrac{3}{5}$ is represented as $\dfrac{6}{10}$, the answer should not change, but now b, such that $b^{10} = (-4)^6$, could be positive. Due to this ambiguity, a calculator will give an error message.

Challenge Set 14

1. $y = 150 \cdot 2^{x/11.9}$

2. $y = 1200 \cdot 2^{x/9}$

3. $y = 4.6 \cdot 2^{x/5.3}$

4. $y = 72.5\left(\dfrac{1}{2}\right)^{x/3.1}$

5. $y = 1450\left(\dfrac{1}{2}\right)^{x/8.3}$

6. $y = 128\left(\dfrac{1}{2}\right)^{x/1.6}$

7. 6.0

8. 4.8

9. 2.9

10. 1.8

11. 3.5

12. 3.8

13. $x > 3.3$

14. $x \geq 4.1$

15. $x > 9.2$

16. a. translated 4 units up

   b. translated 3 units to the right

   c. reflected in the y-axis

17. $y = 25,000(1.06)^x$

18. a. $y = 250\left(1 + \dfrac{r}{100}\right)^{3.5}$

   b. about 10%

Challenge Set 15

1. a. 0.693

   b. $y = e^{-1.44t}$

2. a. 0; as t approaches $-\infty$, e^{-t} gets larger and larger, as does the entire denominator. Thus, y approaches 0.

   b. 0; a; as t approaches $+\infty$, e^{-t} approaches 0, and the denominator gets closer and closer to 1. Thus y approaches a.

3. a. $n = ru; \dfrac{r}{n} = \dfrac{1}{u}; u$ gets large.

   b. $\left(1 + \dfrac{r}{n}\right)^n = \left(1 + \dfrac{1}{u}\right)^{ru} = \left[\left(1 + \dfrac{1}{u}\right)^u\right]^r \to e^r$ as $u \to +\infty$

4. a. 2.71

   b. e^2; e^a

Challenge Set 16

1. a. $y = 0.65(-0.5)^{n-1}$ or $y = -1.3\left(-\dfrac{1}{2}\right)^n$

   b. true from the 4th bounce on

2. As x gets very large, y approaches c. Answers may vary; one possibility is the record time for a track-and-field event by year, where c is the theoretical minimum time with respect to human ability.

3. 3 times; $(-0.77, 0.59)$, $(2, 4)$, $(4, 16)$

4. a. once; $(0, a)$

   b. $ab^x = ac^x \implies b^x = c^x \implies \left(\dfrac{b}{c}\right)^x = 1$; therefore, $x = 0$.

5. a. $y = \dfrac{b^r - 1}{r}x + 1$

   b. $y = \left(\dfrac{b^r - 1}{r}\right)\dfrac{r}{2} + 1 = \dfrac{b^r - 1}{2} + 1 = \dfrac{b^r + 1}{2}$; $b^{r/2}$

   c. $\dfrac{1}{2}(b^{r/2} - 1)^2 = \dfrac{1}{2}(b^r - 2b^{r/2} + 1) =$

   $\dfrac{1}{2}(b^r + 1) - b^{r/2} \geq 0$; therefore,

   $\dfrac{1}{2}(b^r + 1) \geq b^{r/2}$

Chapter 3 Challenge Set

1. a. $\dfrac{5}{6}$

b. $y = N\left(\dfrac{5}{6}\right)^n$; about 9 rolls

2. a. $2^{1.7}$, $2^{1.73}$, $2^{1.732}$; $b^{m/n}$ is defined as $(b^{1/n})^m$; each of these numbers is a rational power of 2 whose exponents can be rewritten as $\dfrac{17}{10}$, $\dfrac{173}{100}$, and $\dfrac{1732}{1000}$, respectively, and thus come under this definition.

b. 8; 4

3. a. $\dfrac{1}{2}$

b. 2.3 in.

4. $(1 + r)^n = \dfrac{M}{M - rP}$; by graphing $y = (1 + r)^x$

(with r given), together with the line

$y = \dfrac{M}{M - rP}$, you can solve for x. Students'

answers will vary depending on their choices for r, P, and M.

Chapter 4

Challenge Set 17

1. a. $f^{-1}(x) = \dfrac{1}{a}x - \dfrac{b}{a}$; they are reciprocals.

b. Both are equal to the identity function.
$f(f^{-1}(x)) = f^{-1}(f(x)) = x$

2. $g(x) = \dfrac{1}{2}x + \dfrac{7}{2}$; $g^{-1}(x) = 2x - 7$

3. a. Every horizontal line intersects the graph at most once.

b. If a horizontal line intersected the graph twice, you would have 2 points on the graph with distinct x-coordinates and the same y-coordinate.

4. 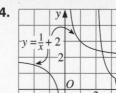 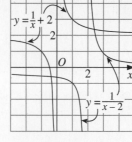 $f^{-1}(x) = \dfrac{1}{x - 2}$

5. 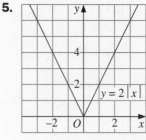 no inverse

6. 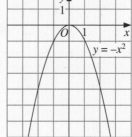 $f^{-1}(x) = x^{1/3}$

7. 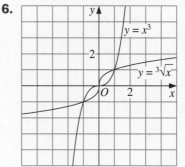 no inverse

8. 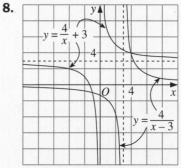 $g^{-1}(x) = \dfrac{4}{x} + 3$

9. 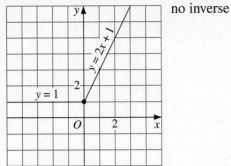 no inverse

10. a. Graphs may vary. A sample is given.

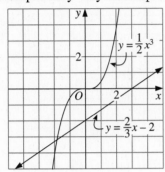

an increasing function

b. Since $f(b) > f(a)$ or $f(a) > f(b)$ whenever a and b are different, the function cannot assign the same y-value to 2 different x-values.

c. Graphs may vary. A sample is given.

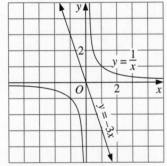

a decreasing function; answer to part (b) is the same as for an increasing function.

11. a.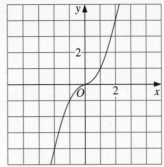

This is an increasing function, so it is one-to-one by the criterion of Exercise 3.

b. $f^{-1}(x) = \sqrt{x}$ if $x \geq 0$, $f^{-1}(x) = -\sqrt{x}$ if $x < 0$.

Challenge Set 18

1. 81

2. –3

3. –4

4. 625

5. 4

6. $\dfrac{1}{2}$

7. $\dfrac{3}{5}$

8. $\dfrac{1}{4}$

9. $\sqrt{5}$

10. 5, –5

11. 6

12. 11

13. **a.** 60

   b. $10^{12} \cdot I_0$

14. **a.** $a \approx \left(1 + \ln \dfrac{a}{n}\right)^n$

   b. $a = n(a^{1/n} - 1)$; about 1.11 (actual value is 1.10)

Challenge Set 19

1. $\dfrac{1}{2}(\log_3 a + 5 \log_3 b) - 3 \log_3 c$

2. $\dfrac{1}{5}(4 \log_3 a + \log_3 b - 4 \log_3 c)$

3. $\log_3 b - \log_3 a - \dfrac{3}{2} \log_3 c$

4. $a + b$

5. $2a$

6. $a - b$

7. $-3b$

8. $a + 2b$

9. $7b$

10. $b + 1$

11. $a + 2$

12. $y = \dfrac{1}{x^3}$

13. $y = \dfrac{9}{2x}$

14. $y = \dfrac{x^2}{2}$

15. $y = \dfrac{5}{x^4}$

16. Since $\log_n x = y$ if and only if $n^y = x$, and since the exponential function is increasing, so is the function $\log_n x$. Every increasing function is one-to-one.

17. $\log(a^{\log b}) = (\log b)(\log a)$, since $\log_b M^k = k \log_b M$. The right-hand side can be reduced to the same expression. Since $\log_n x$ is a one-to-one function and $\log(a^{\log b}) = \log(b^{\log a})$, $a^{\log b} = b^{\log a}$.

18. a. $\log\left[\left(\dfrac{1}{2}\right)\left(\dfrac{2}{3}\right)\left(\dfrac{3}{4}\right)\cdots\left(\dfrac{n-1}{n}\right)\right] = -\log(n + 1)$

b. T_n is a negative number that gets larger and larger in absolute value.

Challenge Set 20

1.

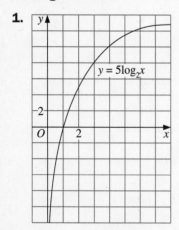

$y = 5\log_2 x$

stretched vertically

2.

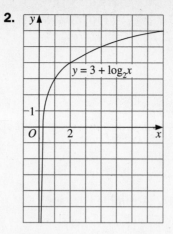

$y = 3 + \log_2 x$

translated 3 units up

3.

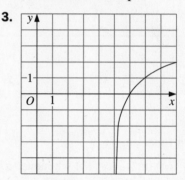

translated 5 units to the right

4. $0 < x < 4$

5. $x \geq 12$

6. $-4 < x < 131$

7. Both sides equal $\log_b M$, by the change-of-base formula.

8. 9

9. 4

10. 0, 2

11. 16, $\dfrac{1}{8}$

12. 256

13. 49

14. 3.419

15. 0.39

16. 16

17. a. $1500(1.005)^n = 1550(1.004)^n$

b. about 33 months

Challenge Set 21

1. $y = 8x^{1/3}$

2. $y = 32(4^{4x})$

3. $y = 15e^{3x}$

4. $y = 6x^2$

5. $y = 8126(1.08)^x$; $y = 645x^{1.09}$; exponential is better.

6. $y = 10{,}476(1.03)^x$; $y = 88x^{1.79}$; power function is better.

7.

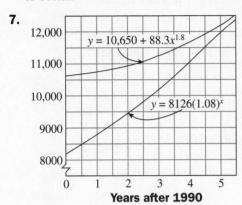

Years after 1990

8. 1996; no

9. The y-values on the graph of $y = \log x$ appear to be larger. Computation shows that for sufficiently large x-values, the y-values on the graph of $y = x^{1/n}$ will be larger.

Chapter 4 Challenge Set

1. a.

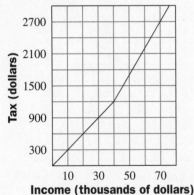

Income (thousands of dollars)

b. The function is increasing, so it is one-to-one; inverse: $f^{-1}(x) = \dfrac{x}{0.03}$ if $0 \le x \le 1200$;

$f^{-1}(x) = \dfrac{x + 800}{0.05}$ if $x > 1200$

2. $x = \log_a b$; $a^x = b$; $a = b^{1/x}$; therefore, $\log_b a = \dfrac{1}{x}$, and $\log_a b \cdot \log_b a = x \cdot \dfrac{1}{x} = 1$.

3. a. $\ln \dfrac{b}{a}$

b. $\ln b$

c. $\ln a$

4. a. By the change-of-base formula,

$\log_{b^2} x = \dfrac{\log_b x}{\log_b b^2} = \dfrac{\log_b x}{2}$; therefore,

$2 \log_{b^2} x = \log_b x$.

b. Answers may vary. A sample is given.
$\log_b x = n\log_{b^n} x$ for any integer n. Proof:

$\log_{b^n} x = \dfrac{\log_b x}{\log_b b^n} = \dfrac{\log_b x}{n}$; therefore,

$n\log_{b^n} x = n\dfrac{\log_b x}{n} = \log_b x.$

5. $\ln (1 + x)^2 = \ln [1 + (2x + x^2)] =$

$(2x + x^2) - \dfrac{(2x + x^2)^2}{2} + \dfrac{(2x + x^2)^3}{3} - \ldots =$

$2x + x^2 - \dfrac{4x^2 + 4x^3 + x^4}{2} + \dfrac{8x^3 + 6x^4 + 12x^5 + x^6}{3} -$

$\ldots = 2x - x^2 + \dfrac{2}{3}x^3 + \ldots =$

$2\left(x - \dfrac{x^2}{2} + \dfrac{x^3}{3} - \ldots\right) = 2 \ln (1 + x)$

Chapter 5

Challenge Set 22

1. $8, -12$

2. $\dfrac{1}{2}, -\dfrac{1}{2}$

3. $\dfrac{1}{3}, -\dfrac{1}{3}$

4. $1.2, -1.2$

5. $-3.2, -4.8$

6. 0

7. If both points are on the parabola, then $q = p^2$ and $2q = 4p^2$. Dividing the second equation by the first results in $2 = 4$, which is a false statement. Thus, both points cannot be on the parabola.

8. a. 3.87

b. 6.08

c. 2.41

d. 0.55

9. 9.44

10. 7.17

11. 8.58

12. 13.09

Challenge Set 23

1. $y = 3(x - 1)^2 + 4$

2. $y = -2(x + 3)^2 + 6$

3. $y = \frac{1}{2}(x - 4)^2 - 3$

4. $y = -\frac{2}{3}(x + 2)^2 - 1$

5. $y = \frac{3}{2}(x + 5)^2 - \frac{1}{2}$

6. $y = 3\left(x - \frac{1}{2}\right)^2 + \frac{1}{4}$

7. $y = 4(x - 1)^2 + 2$

8. $y = \frac{1}{2}(x - 3)^2 - 5$

9. $y = 2(x - a + d)^2 + (b - c)$

10. $y = -\frac{1}{3}(x - 4)^2 + 3$

11. $y = \frac{1}{4}(x - 3)^2 - 2$

12. $x = h \pm \sqrt{\dfrac{y - k}{a}}$; since the line of symmetry is $x = h$, points with these x-coordinates are equidistant from this line.

13. $\dfrac{1}{4a} + k$

Challenge Set 24

1. $y = 2(x - 1)(x + 3)$

2. $y = 3(x + 2)(x + 4)$

3. $y = -5(x - 2)(x - 6)$

4. $y = -\frac{9}{4}(x + 1)(x - 5)$

5. $y = -\frac{1}{4}(x - 3)(x - 7)$

6. $y = -4\left(x - \frac{1}{2}\right)\left(x + \frac{5}{2}\right)$

7. $-4 \le x \le 2$

8. $x < 1$ or $x > 3$

9. $-7 \le x \le -3$

10. $x < -2$ or $x > 5$

11. $y = 2(x + 1)(x - 4)$; $-1, 4$

12. $y = \frac{3}{2}(x - 2)(x - 5)$; $2, 5$

13. $y = -(x + 3)(x - 7)$; $-3, 7$

14. a. In intercept form this equation is $y = (x - r)(x - s)$, where r and s are the solutions. If you multiply the right side of the equation, the constant term is rs.

b. If $y = ax^2 + bx + c$, rewrite the equation as $y = a\left(x^2 + \dfrac{b}{a}x + \dfrac{c}{a}\right) = a(x - r)(x - s)$. The product of the solutions is thus $\dfrac{c}{a}$.

c. The product of the solutions must be $\dfrac{4}{3}$, according to part (b). Therefore, the other solution must be $\dfrac{3}{2}$.

Challenge Set 25

1. a. $-\dfrac{1}{2a}, -\dfrac{1}{4a}$

b.

It is a line.

c. $y = \dfrac{1}{2}x$

2. a. $(x - 12)^2 + (y + 5)^2 = 13^2$; $(12, -5)$

b. $\left(x - \dfrac{1}{2}\right)^2 + \left(y - \dfrac{3}{2}\right)^2 = \dfrac{11}{2}$; $\left(\dfrac{1}{2}, \dfrac{3}{2}\right)$

3. $(x - r)(x - s) = x^2 - (r + s)x + rs = x^2 -$

$(r + s)x + \left(\dfrac{r + s}{2}\right)^2 + rs - \left(\dfrac{r + s}{2}\right)^2 =$

$\left(\dfrac{x - (r + s)}{2}\right)^2 + rs - \dfrac{(r + s)^2}{4}$

4. a. $x^4 + 64 = x^4 + 16x^2 + 64 - 16x^2 =$
$(x^2 + 8)^2 - 16x^2$

b. $(x^2 + 8)^2 - 16x^2 = (x^2 + 8 - 4x)(x^2 + 8 + 4x)$

5. a. $t = \dfrac{x}{v_x};\ y = -16\left(\dfrac{x}{v_x}\right)^2 + v_y\left(\dfrac{x}{v_x}\right);\ h =$

$b - 16\left(\dfrac{x}{v_x}\right)^2.$

b. When $x = a$, the dart's height is

$y = -16\left(\dfrac{a}{v_x}\right)^2 + v_y\left(\dfrac{a}{v_x}\right) = -16\left(\dfrac{a}{v_x}\right)^2 + b$; the

monkey's height is $h = b - 16\left(\dfrac{a}{v_x}\right)^2$; since

these are equal, the dart will hit the monkey.

Challenge Set 26

1. −1.08, 1.55

2. −3.53, 0.85

3. 0.25, 7.91

4. $\dfrac{9}{16}$

5. $\dfrac{4}{3}$

6. $2\sqrt{3}, -2\sqrt{3}$

7. a. $5000(1 + r)^2 + 2500(1 + r)$

b. $5000(1 + r)^2 + 2500(1 + r) = 8268;\ r = 6\%$

8. a. $\dfrac{-b + \sqrt{b^2 - 4ac}}{2a}, \dfrac{-b - \sqrt{b^2 - 4ac}}{2a}$

b. $\dfrac{b}{a}, \dfrac{c}{a}$

9. −2, −1

10. −3, 1

11. $4, -\dfrac{1}{2}$

12. a. $2 - \sqrt{3}; -4, 1$

b. $-2p$

c. Since the sum of the solutions is $-\dfrac{b}{a} = 2p$,
we have $x + p + q\sqrt{r} = 2p$; where x is the
other solution, that is, $x = p - q\sqrt{r}$.

Challenge Set 27

1. −3, 1

2. 2, 1

3. $\ln \dfrac{3}{2}, \ln 5$

4. 3

5. $5, \dfrac{7}{4}$

6. $4, -\dfrac{5}{2}$

7. 0, 8, −2

8. −5, 3, −7

9. $0, \dfrac{1}{2}, -4$

10. −3, 3, −2

11. 5, 3, −3

12. 3, −3, 5, −5

13. $\dfrac{1}{2}, -1$

14. $3, -3, \sqrt{7}, -\sqrt{7}$

15. $(ax + b)(x + 1)$

16. $(x - a)(x + b)$

17. $(ax - 1)(bx - 1)$

18. $(ax + 1)(x - b)$

19. a. The factors must be of the form $(x \pm p)$,
since the coefficient of x^2 is 1. Therefore,
the solutions are integers.

b. $(qx - p)(sx - r)$

c. $a = qs;\ c = pr$

20. 0 must be a solution; $ax^2 + bx = x(ax + b)$

Chapter 5 Challenge Set

1. a. $y = (x - h)^2 + k$

b. $\dfrac{h^2 + k}{2h}$

c. Answers may vary. A sample is given.
$h = 1, k = 7$

2. a. $\dfrac{x}{1} = \dfrac{1 - x}{x};\ x^2 + x - 1 = 0$

b. 0.618

c. The numbers after the decimal point do not
change; the number increases by exactly 1.
The number φ satisfies the equation
$\dfrac{1}{x} = 1 + x.$

3. a. $(250 - 3x)(160 - 2x) = 16,000$

b. 30 ft; there are 2 positive solutions, but the other one, $\frac{400}{3}$ ft is not possible, because twice this length would exceed the rectangle's width.

4. The first differences increase linearly if $a > 0$, decrease linearly if $a < 0$; the second differences are constant and are always equal to $2a$.

Chapter 6

Challenge Set 28

1. categorical

2. a histogram; yes; you could use bars of different colors to represent data from the different countries; a 3-dimensional display could be used, with different countries in rows, one behind another.

3.

Hourly Compensation (Japan)

Hourly Compensation (Mexico)

Hourly Compensation (Norway)

Hourly Compensation (Taiwan)

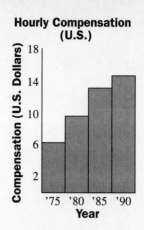

Hourly Compensation (U.S.)

Compensation is increasing fastest in Japan and Taiwan; it is steadiest in the United States; Mexico and Norway have irregular growth.

4. categorical; bar graph

5.

Hourly Compensation (1990)

Compensation rates are vastly different from one country to another.

6. 19.2, 11.2, 8.4, 14.4, 46.8; circle graph

7. Health-related majors, education, and liberal arts have increased; business and accounting and other majors have decreased.

Challenge Set 29

1-11. Answers may vary. Samples are given.

1. Assumption: State legislators work full time; their work may not require them to be on the job 8 hours a day; "Do you think our state legislators should be paid higher wages?"

2. Assumption: Owners can control their dogs; some dogs do not obey their owners' commands; "Do you believe dogs should be allowed in the public parks as long as their owners keep them on a leash?"

3. Assumption: Campaigns will be cleaner; candidates could still get money from other sources; remove the word "cleaner."

Challenge Problems, ALGEBRA 2: EXPLORATIONS AND APPLICATIONS
Copyright © McDougal Littell Inc. All rights reserved.

4. Assumption: Projects will be worthy; some may cater to special interests; "… to finance government projects in other areas?"

5. Assumption: Others have made money by following the newsletter's advice; money may have been made independently of the newsletter; "… newsletter to get information about stocks and bonds."

6. a. Insert "congested" before "downtown."

b. Insert "expensive" before "roads."

7. a. Insert "poor" before "chipmunk."

b. Insert "overcrowded" before "national."

8. a. Insert "just" before "four."

b. Insert "useless" before "county."

9. a. Insert "valuable" before "unmanned."

b. Insert "expensive" before "unmanned."

10. "Would you favor a drive to allow more citizens of the district to exercise their Constitutional right to vote?"

11. "Would you like to eliminate disfranchisement of citizens of the district?"

Challenge Set 30

1. Answers may vary. Samples are given.

a. People who work in stores, who would be out during the day and home in the evening.

b. Municipal workers, who would be out on weekdays and home on weekends.

c. House painters, who would not work on rainy days.

2. a. cluster sample

b. systematic sample

c. random sample

d. convenience sample

3. c; these numbers would be scattered throughout the site.

4. 2 * int(10 * rand + 1)

5. 2 * int(50 * rand + 1) − 1

6. 70 + int(10 * rand)

7. 11 * int(9 * rand + 1)

8. a. Use the nonzero digits of the decimal expansion of the square root of a number that is not a perfect square. Repeat the procedure starting with your previous answer. When zeros appear immediately after the decimal point, eliminate them by multiplying by a power of 10 before taking the next square root.

b. Because a calculator rounds off, a number that is a perfect square could conceivably appear. You could get back to your original number (or 1) this way and then the sequence would repeat.

c. Use the same procedure, but take digits in pairs.

Challenge Set 31

1. a. mode

b. mean

2.

Friday	
Minutes	**Frequency**
2-3	4
4-5	6
6-7	5
8-9	3
10-11	2

Saturday	
Minutes	**Frequency**
4-5	2
6-7	3
8-9	7
10-11	5
12-13	3

3. **Friday 10:00–11:00 A.M. Waiting Times**

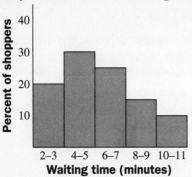

Saturday 4:00–5:00 P.M. Waiting Times

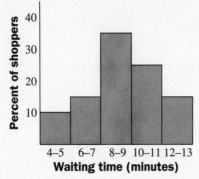

4. Answers may vary. A sample answer is given. The peaks of one histogram would tend to even out the peaks of the other histogram, so the heights of the bars of the new histogram would be more uniform with the exception of the bars at either end.

Waiting Times for Customers

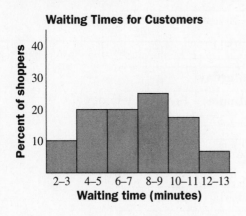

5. a. No; one of the new prices is $\leq \$50$, and one is $\geq \$50$.

b. Yes; both new prices are $\leq \$48$.

6. a.

11	0, 4, 7, 8
12	5, 8
13	1, 2, 3, 3, 5
14	2, 4, 5

b. Answers may vary. A sample is given. A stem-and-leaf plot is essentially a histogram in which the intervals are all 10 and that is oriented horizontally instead of vertically.

Challenge Set 32

1. a. about 0.941

b. $\sqrt{\dfrac{f_1(x_1 - \bar{x})^2 + f_2(x_2 - \bar{x})^2 + \ldots + f_n(x_n - \bar{x})^2}{n}}$,

where $n = f_1 + f_2 + \ldots + f_n$.

2. a. $k\bar{x}$; $k\sigma$

b. $\bar{x} + k$, σ

3. a. $\dfrac{(x_1 - \bar{x})^2 + (x_2 - \bar{x})^2 + \ldots + (x_n - \bar{x})^2}{n} =$

$\dfrac{x_1^2 - 2\bar{x}x_1 + (\bar{x})^2 + x_2^2 - 2\bar{x}x_2 + (\bar{x})^2 + \ldots + x_n^2 - 2\bar{x}x_n + (\bar{x})^2}{n}$

$= \dfrac{x_1^2 + x_2^2 + \ldots + x_n^2}{n} + 2\bar{x}\left(\dfrac{x_1 + x_2 + \ldots + x_n}{n}\right) + \dfrac{n(\bar{x})^2}{n}$

b. $\overline{x^2} - 2(\bar{x})^2 + (\bar{x})^2 = \overline{x^2} - (\bar{x})^2$; $\sigma = \sqrt{\overline{x^2} - (\bar{x})^2}$

4. 6 or 3.5

Challenge 33

1. 9%; 10%

2. 4.1%; 5.3%

3. 180 or 370

4. 256

5. 400

6. a. 10%

b. 5%; 2.5%; the error is halved when the sample is quadrupled.

7. The values are close to 0 in both cases. They are equal. $\sqrt{\dfrac{(1 - \hat{p})(1 - [1 - \hat{p}])}{n}} = \sqrt{\dfrac{\hat{p}(1 - \hat{p})}{n}}$

Chapter 6 Challenge Set

1. a. Answers may vary. A sample is given. $n = 420$, 110, 55, and so on will not produce 20 distinct names; reject values of n that have factors in common with 420.

b. Answers may vary. Samples are given. 1, 22, 79, respectively.

2. 300 or 79

3. AL: $\bar{x} = 2.58$; $\sigma = 0.54$; NL: $\bar{x} = 2.43$; $\sigma = 0.50$

	$x \le \bar{x} - 2\sigma$	$x \le \bar{x} - \sigma$
American	0%	22%
National	4%	12%

	$x < \bar{x}$	$x \le \bar{x} + \sigma$
American	52%	81%
National	58%	88%

	$x \le \bar{x} + 2\sigma$	$x \le \bar{x} + 3\sigma$
American	96%	100%
National	96%	100%

To find the percent of data between $\bar{x} - n\sigma$ and $\bar{x} + n\sigma$, subtract the percent under $\bar{x} - n\sigma$ from the percent under $\bar{x} + n\sigma$; $\bar{x} - \sigma < x \le \bar{x} + \sigma$: 59%; 76%; $\bar{x} - 2\sigma < x \le x + 2\sigma$: 96%; 92%

Chapter 7

Challenge Set 34

1. $\left(\dfrac{1}{3}, -\dfrac{1}{2}\right)$

2. $(\sqrt{5}, \sqrt{2})$, $(\sqrt{5}, -\sqrt{2})$, $(-\sqrt{5}, \sqrt{2})$, $(-\sqrt{5}, -\sqrt{2})$

3. $(4, -10)$, $(4, 4)$, $(-2, -10)$, $(-2, 4)$

4. $(2, 3)$, $(-1, 6)$

5. $(-2, -5)$, $(4, 7)$

6. $(-1, 1)$, $(3, 13)$

7. a. $(3, 2, -1)$

  b. $(-2, 5, 4)$

8. a. $y = \dfrac{7}{15}x - \dfrac{1}{225}x^2$

  b. No; it will hit the crossbar.

9. Solving simulatneously: $x^2 - a^2 = 2ax - 2a^2$; $x^2 - 2ax + a^2 = 0$; $(x - a)^2 = 0$; $x = a$ is the only solution.

Challenge Set 35

1. $\left(\dfrac{21}{5}, -\dfrac{3}{2}\right)$

2. $(\ln 4, \ln 3)$

3. $\left(100, \dfrac{1}{10}\right)$

4. $(-2, 7)$

5. a. $mv_0 = mv + mw$; $\dfrac{1}{2}mv_0^2 = \dfrac{1}{2}mv^2 + \dfrac{1}{2}mw^2$

  b. After cancelling m and $\dfrac{1}{2}$ everywhere and eliminating v_0, you get $(v + w)^2 = v^2 + w^2$; $v^2 + 2vw + v^2 = v^2 + w^2$; $2vw = 0$. The solutions for (v, w) are thus $(v_0, 0)$ and $(0, v_0)$.

  c. If $w = 0$, the second ball would remain at rest, preventing the first ball from having any positive velocity. Therefore, the second solution is correct: the first ball comes to a dead stop, and the second ball moves off with velocity v_0.

6. Multiply the first equation by d, and the second by b, then subtract to get

$adx - bcx = pd - qb$. Thus, $x = \dfrac{pd - qb}{ad - bc} = \dfrac{\det B}{\det A}$.

For y, multiply the first equation by c, the second by a, and use a similar procedure.

Challenge Set 36

1. a. All the other terms contain at least one 0 factor.

  b. $\begin{bmatrix} a & 0 & 0 \\ 0 & e & 0 \\ 0 & 0 & k \end{bmatrix} \begin{bmatrix} ek & 0 & 0 \\ 0 & ak & 0 \\ 0 & 0 & ae \end{bmatrix} =$

  $\begin{bmatrix} aek & 0 & 0 \\ 0 & aek & 0 \\ 0 & 0 & aek \end{bmatrix}$; therefore, $AB = I$. $B = A^{-1}$

2. a. All the other terms contain a 0 factor.

  b. $\begin{bmatrix} a & b & c \\ 0 & e & f \\ 0 & 0 & k \end{bmatrix} \begin{bmatrix} ek & -bk & bf - ce \\ 0 & ak & -af \\ 0 & 0 & ae \end{bmatrix} =$

  $\begin{bmatrix} aek & 0 & 0 \\ 0 & aek & 0 \\ 0 & 0 & aek \end{bmatrix}$; therefore $AB = I$.

3. a. $-1\begin{bmatrix} 1 & 4 & 5 \\ 0 & -1 & -2 \\ 0 & 0 & -1 \end{bmatrix}$

b. $-\dfrac{1}{2}\begin{bmatrix} 2 & -3 & -14 \\ 0 & -1 & -2 \\ 0 & 0 & -2 \end{bmatrix}$

4. a. $(3, -2, 1)$

b. $(2, 5, -4)$

Challenge Set 37

1.

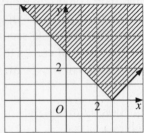

2.

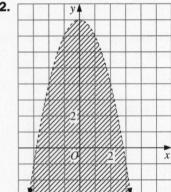

3.

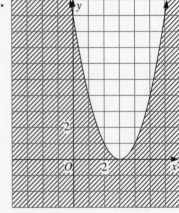

4.

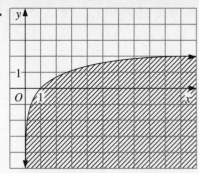

5.

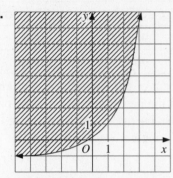

6.

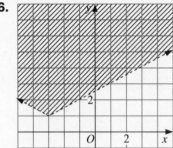

7. $y \ge -\dfrac{3}{2}x + \dfrac{17}{2}$

8. $y < \dfrac{1}{3}x + \dfrac{11}{3}$

9. $y \le \dfrac{2}{5}x - \dfrac{12}{5}$

10. $y > x - 5$

11. a. Answers may vary; for example, the upper half plane including those points $(x, 0)$ such that $x \ge 0$ is one possibility.

b. For the example given in part (a): (x, y) such that $y \ge 0$ and $x \ge 0$ whenever $y = 0$.

12.

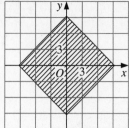

13.

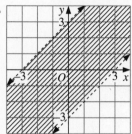

14.

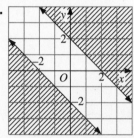

15. a-b.

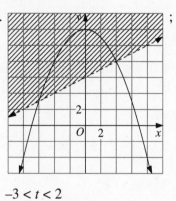

$-3 < t < 2$

16. $y > |60 - 90t|$

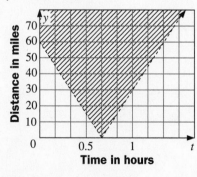

1.

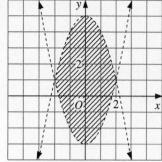

2.

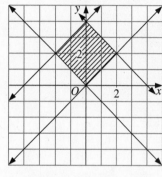

3.

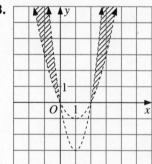

4.

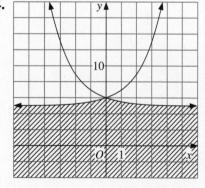

5.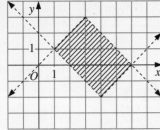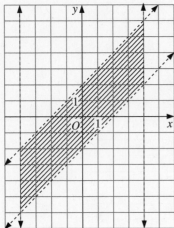

6.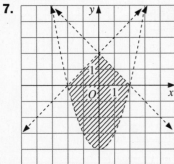

7.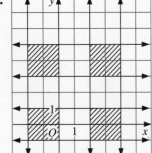

8.

9.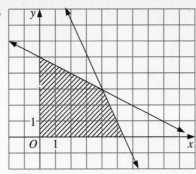

10. $y \geq \frac{2}{5}x + \frac{1}{5}; \; y \leq 2x + 5; \; y \leq -2x + 5$

11. $-1 \leq y \leq 4; \; y \leq \frac{5}{2}x + 4; \; y \leq -\frac{5}{2}x + \frac{23}{2}$

12. $y \geq x - 3; \; y \geq 3 - x; \; y \leq x + 3; \; y \leq 9 - x$

13. $x < 6; \; y < \frac{1}{2}x + 4; \; y > -\frac{1}{2}x + 6$

14. a.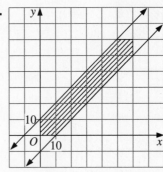

b. $x \geq 0; \; y \geq 0; \; y \leq 60; \; x \leq 60; \; y \leq x + 10; \; y \geq x - 10$

Challenge Set 39

1. a. $x \geq 0; \; y \geq 0; \; \frac{1}{6}x + \frac{1}{3}y \leq 8; \; \frac{1}{3}x + \frac{1}{3}y \leq 9; \; \frac{1}{2}x + \frac{1}{8}y \leq 9$

b.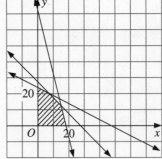

c. $(15, 12)$

2. If $P(a, b)$ is an interior point, there is a point near P of the form $(a + c, b)$ at which the function $x + y$ has the value $a + c + b > a + b$. Therefore, $x + y$ is not maximized at (a, b).

3. a. If the line $x + y = k$ were not tangent to the parabola, it would pass through an interior point, and the argument in Exercise 2 shows that k would not be a maximum.

b. $x \geq 0$; $y \geq 0$; $y \leq 4 - x^2$

c. Since the slope of $x + y = k$ is -1, $-2a = -1$ and thus $a = \dfrac{1}{2}$; $\left(\dfrac{1}{2}, \dfrac{15}{4}\right)$; $\dfrac{17}{4}$

Chapter 7 Challenge Set

1. a. $x - y$

b. $1 + \dfrac{x - y + 1}{x - y}$

c. $\dfrac{1}{y}$; $y = \dfrac{1}{2}$; no

2. If the vertices of the original square are $A(0, 0)$, $B(1, 0)$, $C(1, 1)$, and $D(0, 1)$, then the images are A', B', C', and D' as shown in the diagrams.

a. The image is the same as the original; points are reflected across the line $y = x$.

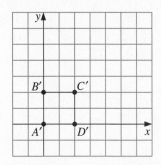

b. The image is the parallelogram with vertices $(0, 0)$, $(1, 0)$, $(1, 2)$, and $(1, 1)$; points are "sheared" to the right proportionally to their height above or below the x-axis.

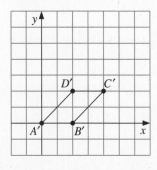

c. The image is the square with vertices $(0, 0)$, $(-1, 0)$, $(-1, 1)$, and $(0, 1)$; points are reflected across the y-axis.

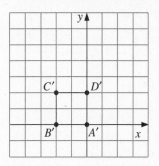

d. The image is the square with the same vertices as in part (c); points are reflected across the line $y = x$, then reflected across the y-axis.

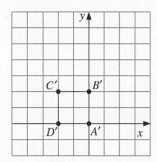

3.

4. $(9, 0)$: 27; $(7, 5)$: 19; the segment joining $(4, 8)$ and $(7, 5)$: 12; $(4, 8)$: 44; $(0, 10)$: 40. The function $ax + by$ will have its maximum at a corner point if and only if the slope $-\dfrac{a}{b}$ of the lines $ax + by = k$ is between the slopes of the sides of the polygon that meet at that corner point. If $-\dfrac{a}{b}$ equals the slope of a side, every point on that side will be a maximum.

Challenge Set 40

1. $f^{-1}(x) = (x + 3)^2,\ x \geq -3$

2. $f^{-1}(x) = -\sqrt{x - 1}$

3. $f^{-1}(x) = (x - 3)^2 - 5,\ x \leq 3$

4. $f^{-1}(x) = \sqrt{x - 1} + 4$

5. $f^{-1}(x) = \frac{1}{2}[(x + 3)^2 + 7],\ x \geq -3$

6. $f^{-1}(x) = \sqrt{\dfrac{x + 4}{2}} - 3$

7. If $x \geq 0$, then $\sqrt{x^2} = x$; if $x < 0$, $\sqrt{x^2} = -x$.

8. a. $5|x|\sqrt{2}$

    b. $4y\sqrt{3y}$

    c. $10|q|\sqrt{5p}$

    d. $7a^2\sqrt{10}$

9. $x \geq 19$

10. $-4 \leq x \leq 5$

11. $4 \leq x \leq \dfrac{41}{4}$

12.

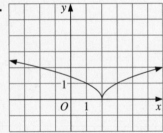

D: all real numbers, R: all real numbers ≥ 0

13.

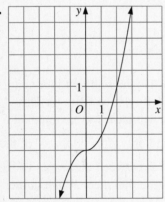

D: all real numbers, R: all real numbers

14.

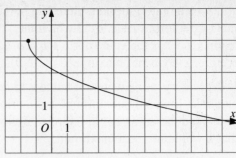

D: all real numbers $\geq -\dfrac{3}{2}$, R: all real numbers ≤ 5

15. a. Answers may vary. A sample is given.
0 and 1

    b. At least one of the two numbers must be 0; if $\sqrt{a} + \sqrt{b} = \sqrt{a + b}$, then $a + 2\sqrt{ab} + b = a + b$; hence, $2\sqrt{ab} = 0$; therefore a or b must $= 0$.

16. $(3, -2)$

17. a. domain: $x \geq -\dfrac{b}{a}$; range: $y \geq c$

    b. $f^{-1}(x) = \dfrac{1}{a}[(x - c)^2 - b],\ x \geq c$

    c. The vertex of the graph of f^{-1} is at $\left(c, -\dfrac{b}{a}\right)$; therefore, the vertex of the graph of f is at $\left(-\dfrac{b}{a}, c\right)$.

Challenge Set 41

1. $2x\sqrt[3]{3}$

2. $2a\sqrt[5]{3a^4}$

3. $6y^2$

4. $\dfrac{u}{2v}$

5. $6x^2\sqrt[6]{5}$

6. $\dfrac{q^2}{3p}$

7. a. 2

    b. $\sqrt[m]{\sqrt[n]{x}} = (\sqrt[n]{x})^{1/m} = (x^{1/n})^{1/m} = x^{1/mn} = \sqrt[mn]{x}$

8.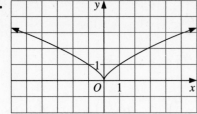

9.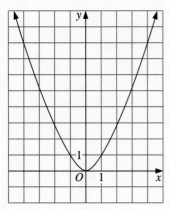

10.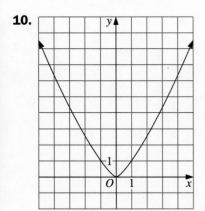

11.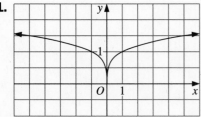

12. Graphs in Exercises 8 and 11 come to a point at the origin; graphs in Exercises 9 and 10 do not. If $\frac{p}{q} < 1$ in the function $y = \left| x \right|^{p/q}$, the graph will come to a point; if $\frac{p}{q} > 1$, the graph will be smooth.

13. a. Both of these mean the product of x with itself pq times, since $pq = qp$.

b. $y^n = \left(\sqrt[n]{x^m} \right)^n = x^m;$

$z^n = \left[\left(\sqrt[n]{x} \right)^m \right]^n = \left(\sqrt[n]{x} \right)^{mn} = \left(\sqrt[n]{x} \right)^{nm} =$

$\left[\left(\sqrt[n]{x} \right)^n \right]^m = x^m;$ therefore, $y^n = z^n$, so $y = z$,

since both are positive, and every power function is one-to-one on the positive numbers.

14. a.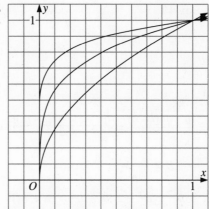

b. The point at the origin remains fixed, and the rest of the graph gets closer and closer to the line $y = 1$. $\sqrt[n]{a}$ approaches 1 for all a, $0 < a \leq 1$.

Challenge Set 42

1. 16

2. $\dfrac{25}{4}$

3. $\dfrac{9}{4}$

4. 9

5. $-1, -6$

6. 11

7. 17

8. $\dfrac{17}{16}$

9. 4, 12

10. -2

11. 3, 11

12. 14, -2

13. 5

14. 3, –1

15. 9

16. $\frac{1}{2}, \frac{9}{8}$

17. 1, –2

18. 0, 2

19. a. $\dfrac{\sqrt{x^2 + 0.36}}{2}; \dfrac{2 - x}{6}$

   b. 0.8 mi

Challenge Set 43

1. $3i$

2. $-2i, -i$

3. $-3i, \dfrac{4}{3}i$

4. $\dfrac{p + qi}{r + si} = \dfrac{(p + qi)(r - si)}{r^2 + s^2} = \dfrac{(pr + qs) + (qr - ps)i}{r^2 + s^2};$

   $a = \dfrac{pr + qs}{r^2 + s^2}; b = \dfrac{qr - ps}{r^2 + s^2}$

5. a. $(-a - bi)^2 = a^2 + 2abi - b^2 = (a + bi)^2$

   b. $(4 - 3i)^2 = 16 - 24i - 9 = 7 - 24i$

   c. $-4 + 3i$

6. $-i, 1, i, -1; i^n = i^r$

7. a. $\bar{z} + \bar{w} = (a - bi) + (c - di) =$
   $(a + c) - (b + d)i = \overline{z + w}$

   b. $\bar{z} \cdot \bar{w} = (a - bi)(c - di) =$
   $(ac - bd) - (ad + bc)i = \overline{z \cdot w}$

   c. $k \cdot \bar{z} = k(a - bi) = ka - kbi = \overline{kz}$; This is the
   same as $\bar{k} \cdot \bar{z} = \overline{kz}$, since $k = \bar{k}$ for all real
   numbers k.

8. a. Each term of $P(z)$ has the form az^n, so by
   repeated application of parts (b) and (c) of
   Exercise 7, $\overline{az^n} = \bar{a} \cdot (\bar{z})^n$. By part (a) of
   Exercise 7, therefore, $P(\bar{z}) = \overline{P(z)}$.

   b. If $P(z) = 0$, then $\overline{P(z)} = \bar{0}$; that is, $\overline{P(z)} = 0$.
   Therefore, $P(\bar{z}) = 0$, by part (a).

9. a. $\left[\dfrac{1}{2}\left(-1 + i\sqrt{3}\right)\right]^3 =$

   $\dfrac{1}{8}\left(-1 + i\sqrt{3}\right)\left(-1 + i\sqrt{3}\right)^2 =$

   $\dfrac{1}{8}\left(-1 + i\sqrt{3}\right)\left(-2 - 2i\sqrt{3}\right) = \dfrac{1}{8}(2 + 6 + 0i) =$

   1; therefore $z^3 - 1 = 0$.

   b. $\dfrac{1}{2}\left(-1 - i\sqrt{3}\right); \left[\dfrac{1}{2}\left(-1 - i\sqrt{3}\right)\right]^3 =$

   $\dfrac{1}{8}\left(-1 - i\sqrt{3}\right)\left(-1 - i\sqrt{3}\right)^2 =$

   $\dfrac{1}{8}\left(-1 - i\sqrt{3}\right)\left(-2 + 2i\sqrt{3}\right) =$

   $\dfrac{1}{8}(2 + 6 + 0i) = 1$

Challenge Set 44

1. a.

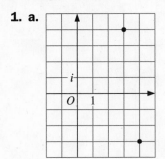

   b.

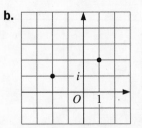

   c.

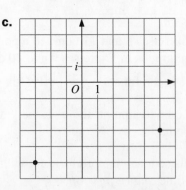

d.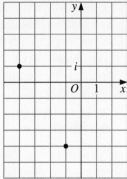

2. The segments have the same length and form a right angle; multiplying by $-i$ rotates a complex number 90° clockwise.

3. a.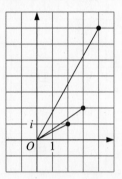

b.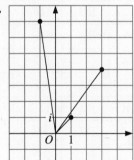

c. 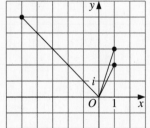 ;

The angle of the product is the sum of the angles of the factors.

4. The magnitude of the product is the product of the magnitudes of the factors.

5. The angle of the quotient will be the difference of the angles of the dividend and divisor; the magnitude of the quotient will be the quotient of the magnitudes.

6.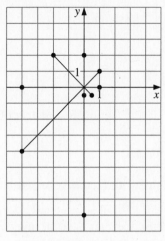

a. The angles increase by 45°.

b. The magnitudes are $(\sqrt{2})^n$.

c. see graph above; yes

d. $z^6 = -8i$; see graph above; yes

7. 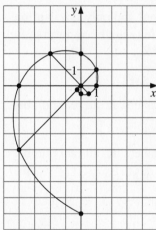 The radial lines all make the same angle with the spiral.

Challenge 45

1. a. $IJ = J$, since I is an identity, and $IJ = I$, since J is an identity. Therefore, $I = J$.

b. $C = IC = (BA)C = B(AC) = BI = B$

2. a. Yes, since the set $\{1, -1\}$ is closed under multiplication.

b. $\begin{bmatrix} 1 & 0 \\ 0 & 1 \end{bmatrix}$

---

c. No; answers may vary. A sample is given.

$$\begin{bmatrix} 2 & 5 \\ 1 & 3 \end{bmatrix}\begin{bmatrix} 2 & 1 \\ 7 & 4 \end{bmatrix} = \begin{bmatrix} 39 & 22 \\ 23 & 13 \end{bmatrix}, \text{ but}$$

$$\begin{bmatrix} 2 & 1 \\ 7 & 4 \end{bmatrix}\begin{bmatrix} 2 & 5 \\ 1 & 3 \end{bmatrix} = \begin{bmatrix} 5 & 13 \\ 18 & 47 \end{bmatrix}$$

3. a. Yes; $\begin{bmatrix} a & -b \\ b & a \end{bmatrix}\begin{bmatrix} c & -d \\ d & c \end{bmatrix} =$

$\begin{bmatrix} ac - bd & -(ad + bc) \\ ad + bc & ac - bd \end{bmatrix}$; this is a matrix of

the required form.

b. Yes; see part (a); with factors reversed, the product is the same.

c. No; the 0 matrix does not have an inverse; every other such matrix does.

d. The product of corresponding matrices is the matrix corresponding to the product of the complex numbers; the reciprocal of a complex number corresponds to the inverse of the corresponding matrix.

4. No, Z_n is a group if and only if n is prime; otherwise, there may not be an inverse for every element of Z_n.

Chapter 8 Challenge Set

1. a. D: all real numbers, R: all real numbers; $f^{-1}(x) = \sqrt[3]{x - 1}$

b. D: all real numbers, R: all real numbers ≥ 0; restrict to $x \geq 0$; $f^{-1}(x) = \sqrt[3]{x}$

c. D: all real numbers > 3, R: all real numbers > 2; $f^{-1}(x) = 3 + \dfrac{1}{(x-2)^2}$, $x \geq 2$

2. a. $(i^n z)^4 = (i^4)^n z^4 = 1^n \cdot i = i$; $z = p + qi$; $iz = -q + pi$; $i^2 z = -p - qi$; $i^3 z = q - pi$

b. $(p + qi)^2 = \dfrac{\sqrt{2}}{2} + \left(\dfrac{\sqrt{2}}{2}\right)i$; and $\left[\dfrac{\sqrt{2}}{2} + \left(\dfrac{\sqrt{2}}{2}\right)i\right]^2 = i$

3. Since the sum is real, its imaginary part $b + d = 0$; therefore $d = -b$. Since the product is real, its imaginary part $ad + bc = 0$; that is, $ad - cd = 0$. Since $d \neq 0$, this implies that $a = c$; $c + di = a - bi$, the conjugate of $a + bi$.

4. 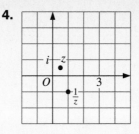 These two points are on radial lines that make equal angles with the real axis; one is a real multiple of the conjugate of the other; their magnitudes are reciprocals of each other.

5.

×	I	R1	R2
No change = I	I	R1	R2
90° CC Rotation = R1	R1	R2	R3
180° Rotation = R2	R2	R3	I
270° CC Rotation = R3	R3	I	R1
Horizontal flip = HF	HF	D1	VF
Vertical flip = VF	VF	D2	HF
Flip about UL to LR diag. = D1	D1	VF	D2
Flip about LL to UR diag. = D2	D2	HF	D1

×	R3	HF	VF
No change = I	R3	HF	VF
90° CC Rotation = R1	I	D2	D1
180° Rotation = R2	R1	VF	HF
270° CC Rotation = R3	R2	D1	D2
Horizontal flip = HF	D2	I	R2
Vertical flip = VF	D1	R2	I
Flip about UL to LR diag. = D1	HF	R3	R1
Flip about LL to UR diag. = D2	VF	R1	R3

×	D1	D2	
No change = I	D1	D2	
90° CC Rotation = R1	HF	VF	
180° Rotation = R2	D2	D1	
270° CC Rotation = R3	VF	HF	
Horizontal flip = HF	R1	R3	
Vertical flip = VF	R3	R1	
Flip about UL to LR diag. = D1	I	R2	
Flip about LL to UR diag. = D2	R2	I	

The symmetries form a noncommutative group; subgroups: {I, R1, R2, R3}, {I, R2}, {I, HF}, {I, VF}, {I, D1}, {I, D2}

Chapter 9

Challenge 46

1. $3i$

2. $-11 + 29i$

3. 0

4. $2\sqrt{3} - 8 + (1 + 8\sqrt{3})i$

5. a. Answers may vary. A sample is given. $y = x^2 - 4$ and $y = 3x^2 - 12$ both pass through $(2, 0)$ and $(-2, 0)$.

  b. 3; you get a system of three equations in three unknowns, which typically has a unique solution.

6. a. $3a + b, 5a + b, 7a + b; 2a, 2a$

  b. $2, -5, 4$

7. a. $7a + 3b + c, 19a + 5b + c, 37a + 7b + c,$ $61a + 9b + c, 91a + 11b + c; 12a + 2b;$ $18a + 2b, 24a + 2, 30a + 2b; 6a, 6a, 6a;$ $0, 0$

  b. $1, -3, -2, 8$

8. $3x^2 - 3x - 4$

9. a. $3.24, 3.115, 3.02, 3.01; 3;$ no

  b. The value is the leading coefficient a.

Challenge 47

1. $x^3 + 3ax^2 + 3a^2x + a^3$

2. $x^4 - 4ax^3 + 6a^2x^2 - 4a^3x + a^4$

3. $a^3x^3 + 3a^2bx^2 + 3a^2x + b^3$

4. a. Each term $(x)(a^jx^{n-j})$ of the product has a corresponding term $(-a)(a^{j-1}x^{n-j+1})$ that cancels it, except for the terms $(x)(x^n)$ and $(-a)(a^n)$.

  b. $(x - 2)(x^4 + 2x^3 + 4x^2 + 8x + 16)$

5. a. Each term $(x)(-a^jx^{n-j})$ of the product has a corresponding term $(a)(a^{j-1}x^{n-j+1})$ that cancels it, except for the terms $(x)(x^n)$ and $(a)(-a^n)$.

  b. $(x + 10)(x^4 - 10x^3 + 100x^2 - 1000x^3 + 10{,}000x^4)$

6. $x + (3 + i) + \frac{-2 + 3i}{x - i}$

7. $4x^2 + (1 - 2i)x - \frac{1 + i}{2} + \frac{-11 + i}{4x + 2i}$

8. $-ix^2 + x - 1 - \frac{x}{ix + 1}$

9. a. $\left(u^3 - bu^2 + \frac{b^2}{3}u - \frac{b^3}{27}\right) + b\left(u^2 - \frac{2bu}{3} + \frac{b^2}{9}\right) +$ $c\left(u - \frac{b}{3}\right) + d = u^3 + \left(c - \frac{b^2}{3}\right)u + \frac{2b^3 - 9bc}{27} + d$

  b. Calculate $u - \frac{b}{3} = x$.

10. a. 8, 8

  b. 4

11. a. $1 - x + x^2 - x^3 + \ldots$

  b. $1 + x^2 + x^4 + x^6 + \ldots = 1 + x^2 + (x^2)^2 +$ $(x^2)^3 + \ldots$ This makes sense, since the product is $\frac{1}{1 - x^2}$, and the power series is the same as the one for $\frac{1}{1 - x}$ with x replaced by x^2.

Challenge 48

1. a. The end behaviors on the two ends of an odd-degree polynomial must be opposite in sign.

  b. Since the graph of an odd-degree polynomial must contain points with arbitrarily large positive y-coordinates, as well as points with y-coordinates that are negative and arbitrarily large in absolute value, such a graph must cross the x-axis.

c. If (a, b) were a maximum, there would be no points (x, y) on the graph with $y > b$, but this is not true, for the reason given in part (b). A similar argument works for minimums.

5. a. $r = \sqrt{100 - x^2}; \ 0 \le x \le 10$

b. $V = \frac{1}{3}\pi(100 - x^2)(10 + x)$

2. a.

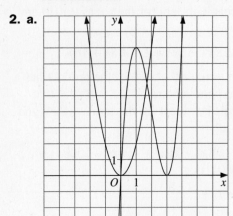

c. $\frac{10}{3}; \frac{40}{3}$

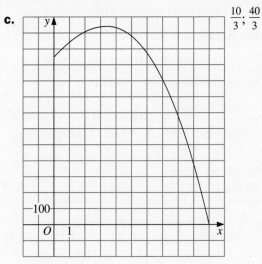

b. The graph of f near $x = 0$ is similar to the graph of g near $x = 3$.

3. a.

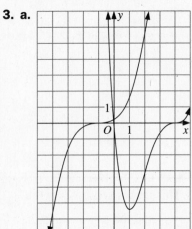

b. The graph of f near $x = -1$ is similar to the graph of g near $x = 3$.

4. Answers may vary. A sample is given. The graph of an odd-degree polynomial must intersect the x-axis an odd number of times; the graph of an even-degree polynomial must intersect the x-axis an even number of times (possibly 0). This is related to the fact that an odd-degree polynomial must have as many local maximums as local minimums; an even-degree polynomial has one more local maximum than it has local minimums, or vice versa.

Challenge Set 49

1. a.

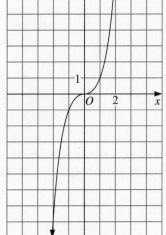

Challenge Problems, ALGEBRA 2: EXPLORATIONS AND APPLICATIONS
Copyright © McDougal Littell Inc. All rights reserved.

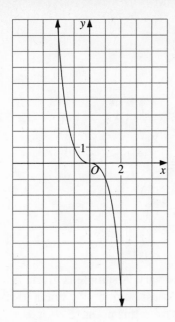

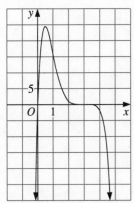

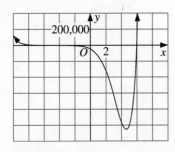

b. Near this point, the graph resembles the graph of $y = x^3$ or $y = -x^3$ near 0.

2. The graph near an even zero resembles the graph of $y = x^2$ or $y = -x^2$.

3. 5, –2

4. The constant term is $-pqr$; that is, the negative of the product of the zeros (by the Factor Theorem). For fourth-degree polynomials, the

theorem says that the constant term equals the product of the zeros (no negative sign). It is proved likewise, by multiplying out the factored form. In general, the constant term is the product of the zeros for an even-degree polynomial and the negative of this product for an odd-degree polynomial.

5. a. $-d = ar^3 + br^2 + cr = r(ar^2 + br + c)$; since a, b, c, and r are all integers, $r(ar^2 + br + c)$ is an integer, and so r is a factor of both sides of the equation.

b. r must be a factor of the constant term d.

6. a. $P(x) = (x - k)Q(x) + R$; since degree of $R(x)$ < degree of $x - k$, R must be a constant.

b. Upon substituting k for x in the given equation, we obtain $P(k) = (k - k)Q(k) + R$; that is, $P(k) = R$. Therefore, $P(k) = 0$ if and only if $R = 0$.

Challenge Set 50

1. The only possible rational zeros of $x^2 - 5$ are ± 1 and ± 5.

2. The only possible rational zeros of $x^3 - 2$ are ± 1 and ± 2.

3. The only possible rational zeros of $x^5 - 6$ are ± 1, ± 2, ± 3, and ± 6.

4. The only possible rational zeros of $3x^2 - 2$ are ± 1, ± 2, $\pm\frac{1}{3}$, and $\pm\frac{2}{3}$.

5. The only possible rational zeros of f are ± 1 and $\pm\frac{1}{2}$, so f has no rational zeros. But since $f(2) = 7 > 0$ and $f(1) = -2 < 0$, the graph of f must cross the x-axis between 1 and 2.

6. a. 3

b. 4

7. $f(k) = (k - k)Q(k) + R = R$

8. a. $a\left(\dfrac{p}{q}\right)^3 + b\left(\dfrac{p}{q}\right)^2 + c\left(\dfrac{p}{q}\right) + d = 0$; $ap^3 + bp^2q + cpq^2 + dq^3 = 0$; $ap^3 = -(bp^2q + cpq^2 + dq^3)$ $= -q(bp^2 + cpq + dq^2)$. Therefore, q is a factor of ap^3; since p and q have no common factor, q must be a factor of a.

b. $a\left(\dfrac{p}{q}\right)^3 + b\left(\dfrac{p}{q}\right)^2 + c\left(\dfrac{p}{q}\right) + d = 0;\ a + b\left(\dfrac{q}{p}\right) +$

$c\left(\dfrac{q}{p}\right)^2 + d\left(\dfrac{q}{p}\right)^3 = 0;$ this is the same as the

original equation with $\dfrac{q}{p}$ in place of $\dfrac{p}{q}$ and

with the roles of a and d reversed.
Therefore, the same argument shows that p
must divide d.

Challenge Set 51

1. a. $\dfrac{1}{x} \le \dfrac{1}{y}$; multiply both sides by the positive

number xy: $y \le x$.

b. If x and y are negative, the same inequality
holds. The same proof works (since xy is
still positive). If x is negative and y is
positive, then $x \le y$.

2. $0 < x < 4$

3. $a > -1$

4. $n > 3$

5. a. 67 lb

b. 36,878 mi

6. a. $\ln ab = \ln a + \ln b;\ \ln a + \ln b$

b. $A\dfrac{ab}{a} = A\dfrac{b}{1}$

7. 0.4916, 0.5311, 0.5456; 0.5772

Challenge Set 52

1. $y = \dfrac{3 - x}{x - 1}$

2. $y = \dfrac{5 - 2x}{1 - 4x}$

3. $y = \dfrac{2 - x}{x + 7}$

4. $y = \dfrac{a}{(x - k) + h}$

5. a. $y = \pm(x - h) + k$

b. If $a > 0$, the line $y = x - h + k$ intersects the
hyperbola; if $a < 0$, the line $y = -(x - h) + k$
does.

c. $(\sqrt{a} + h,\ \sqrt{a} + k),\ (\sqrt{a} + h,\ -\sqrt{a} + k),$
$(-\sqrt{a} + h,\ \sqrt{a} + k),\ (-\sqrt{a} + h,\ -\sqrt{a} + k)$

6. $y = \dfrac{3}{x - 2} + 4$

7. $y = \dfrac{10}{x + 5} - 1$

8.

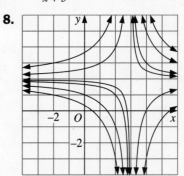

Answers may vary. A sample is given. As $|a|$
grows, the closest point on the graph to (2, 3)
moves away from this point.

9. a.

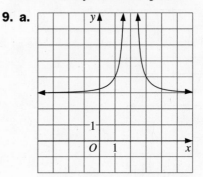

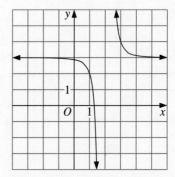

b. The graph of the first equation lies in
Quadrants I and II (relative to the point
(2, 3)) rather than Quadrants I and III, as in
the case of the hyperbola. The graph of the
second equation lies in the same quadrants
as the hyperbola.

c.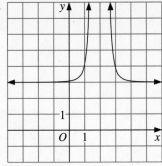

Challenge Set 53

1.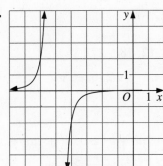

2.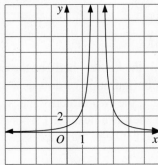

3.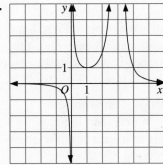

4. Near a zero of odd degree, the two branches of the graph go off in opposite directions, one branch to $+\infty$, one branch to $-\infty$. Near a zero of even degree, both branches go off in the same direction.

5. $\dfrac{-5}{x-1} + \dfrac{2}{x-4}$

6. $\dfrac{3}{x-3} + \dfrac{4}{x+3}$

7. $\dfrac{7}{x+3} - \dfrac{3}{x+1}$

8. $y = \dfrac{8}{x^2+4}$; by slopes, $\dfrac{2}{x} = \dfrac{8}{a(x^2+4)}$. Solving for a, you get $a = \dfrac{4x}{x^2+4}$. Substituting (a, y), the equation of the circle $a^2 + (y-1)^2 = \dfrac{16x^2}{(x^2+4)^2} + \dfrac{(4-x^2)^2}{(x^2+4)^2}$. The right-hand side can be simplified to 1.

Challenge Set 54

1. 4

2. -2

3. $-1, 14$

4. $1, \dfrac{7}{2}$

5. $\dfrac{1}{3}$

6. -2

7. $\dfrac{1}{2}, 5$

8. $1, -\dfrac{1}{2}$

9. Multiplying through by $2p^2 - p$, you get $8 + 8p - 8 = 3p$; $p = 0$; this solution is extraneous.

10. Both of the solutions, 1 and 2, are extraneous.

11. Multiplying through by $x(x+2)(x+3)$, you get, after simplification, $12 = 0$.

12. The only solution, 3, is extraneous.

13. $\dfrac{k}{2}, -\dfrac{k}{4}$

14. a. $\dfrac{d}{2-3c}$

   b. No restrictions other than if $c = \dfrac{2}{3}$, $d = 0$; there is no unique solution whenever $d = 0$.

15. 35 mi/h

Chapter 9 Challenge Set

1. a. When x_j is substituted for x, two of the terms are 0, and the other is $y_j \cdot 1$. Each term has degree 2; hence, so does $P(x)$.

b. $y_0 \dfrac{(x-x_1)(x-x_2)(x-x_3)}{(x_0-x_1)(x_0-x_2)(x_0-x_3)} + \ldots +$

$y_3 \dfrac{(x-x_0)(x-x_1)(x-x_2)}{(x_3-x_0)(x_3-x_1)(x_3-x_2)}$ (4 terms)

2. a.

Pos.	Neg.	Imag.
3	2	0
3	0	2
1	2	2
1	0	4

b.

Pos.	Neg.	Imag.
3	3	0
3	1	2
1	3	2
1	1	4

3. a. $\dfrac{y}{x}$

b. $\dfrac{a-b}{a+b}$

4. a. A zero is between $2\frac{23}{32}$ and $2\frac{3}{4}$.

b. A zero is between $\frac{13}{32}$ and $\frac{7}{16}$.

Chapter 10

Challenge Set 55

1. 37, 50, 65, 82

2. $\sqrt{2}, \sqrt[4]{2}, \sqrt[8]{2}, \sqrt[16]{2}$

3. S, S, E, N

4. $7, \frac{1}{8}, 9, \frac{1}{10}$

5. $t_n = 1 + \dfrac{(-1)^n}{n} + 1$; $\frac{5}{6}, \frac{8}{7}, \frac{7}{8}$

6. $t_n = 7 + 7(10^{-n})$; 7.00007, 7.000007, 7.0000007

7.

8. 50; $t_n = 2n^2$

9. 14, 20;

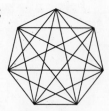 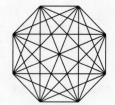

10. a. 1, 7, 19, 37, 61

b. 1, 7, 19, 37, 61

c. They are equal; the "outer shell" of the cube has the same configuration of dots as a hexagon (in fact, the two-dimensional picture of this shell is a hexagon). This shell is the difference between two successive cubes of dots.

Challenge Set 56

1. a. 8, −32

b. 12, 18, 27 or −12, 18, −27

c. $3\sqrt{2}, 6, 6\sqrt{2}$ or $-3\sqrt{2}, 6, -6\sqrt{2}$

d. $\dfrac{1}{5}, \dfrac{1}{2}, \dfrac{5}{4}, \dfrac{25}{8}$

2. $(a-b)^2 \geq 0$; $a^2 - 2ab + b^2 \geq 0$; $a^2 + 2ab + b^2 \geq 4ab$; $a + b \geq 2\sqrt{ab}$; $\dfrac{a+b}{2} \geq \sqrt{ab}$; yes, if $a = b$.

3. CD is the geometric mean between AD and DB; CE is their arithmetic mean; since the perpendicular distance is the shortest distance between a point and a line, $CD \leq CE$.

4. The sequence of logs is an arithmetic sequence; write $t_n = ar^n$, then $\log t_n = \log a + n\log r$, this sequence is arithmetic with common difference $\log r$.

5. The sequence 2^{a_n} is geometric, since $2^{a+nd} = 2^a \cdot (2^d)^n$, which is the general term of a geometric sequence with common ratio 2^d.

6. $c = a + 2x$; since the sequence is geometric, we must have $b^2 = ac$; that is, $(a + x)^2 = a(a + 2x)$; $a^2 + 2ax + x^2 = a^2 + 2ax$; hence, $x = 0$.

7. $1, 31, 61, 91, \ldots$

8. a. -1

   b. $3, -2$

Challenge Set 57

1. $t_1 = a$; $t_n = a(t_{n-1} + 1)$

2. $t_1 = a$; $t_n = bt_{n-1} - a$

3. $t_1 = 1$; $t_n = n \cdot t_{n-1}$

4. a. $1, 4, 9, 16, 25$

   b. $t_n = n^2$

   c. $t_k = (k-1)^2 + 2\sqrt{(k-1)^2} + 1 =$
      $k^2 - 2k + 1 + 2(k-1) + 1 = k^2$

5. a. $2, 6, 12, 20, 30$

   b. $t_n = t_{n-1} + 2n$

   c. $t_n = n^2 + n$

6. a. ; $1, 5, 21, 85$

   b. $t_n = 4t_{n-1} + 1$

Challenge Set 58

1. 7500

2. a. $n(n + 1)$

   b. 2450

   c. 30

   d. $\dfrac{pn(n+1)}{2}$

3. $12a + 78b$

4. $12a - 72b$

5. a. $\dfrac{n(n+1)}{2}$

   b. $\dfrac{1(1+1)}{2} = 1$; $1 + 2 + 3 + \ldots + k = \dfrac{k(k+1)}{2}$;

   $1 + 2 + 3 + \ldots + k + (k+1) =$

   $\dfrac{k(k+1)}{2} + (k+1) = \dfrac{k(k+1) + 2(k+1)}{2} =$

   $\dfrac{(k+1)(k+2)}{2} = (k+1)\dfrac{(k+1)+1}{2}$

6. $\dfrac{1(1+1)(2+1)}{6} = \dfrac{6}{6} = 1$; $1^2 + 2^2 + 3^2 + \ldots + n^2 +$

   $(n+1)^2 = \dfrac{n(n+1)(2n+1) + 6(n+1)^2}{6} =$

   $(n+1)\dfrac{n(2n+1) + 6(n+1)}{6} =$

   $(n+1)\dfrac{2n^2 + 7n + 6}{6} = (n+1)(n+2)\dfrac{2n+3}{6} =$

   $(n+1)[(n+1)+1]\dfrac{2(n+1)+1}{6}$

7. 9170

Challenge Set 59

1. a. $a = r = \dfrac{1}{2}$; $S_n = \left(\dfrac{1}{2}\right)\dfrac{1 - \left(\frac{1}{2}\right)^n}{1 - \frac{1}{2}} = 1 - \left(\dfrac{1}{2}\right)^n$

   b. No; $S_n = \left(\dfrac{1}{p-1}\right)\left(1 - \dfrac{1}{p^n}\right)$

2. a. $1 - x^n$

   b. $a + ar + ar^2 + \ldots + ar^{n-1} =$

   $a(1 + r + r^2 + \ldots + r^{n-1}) = \dfrac{a(1 - r^n)}{1 - r}$
   by part (a).

3. $(b + 1)^{n+1} - 1$

4. $1 - \left(\dfrac{a-1}{a}\right)^n$

5. a. $\dfrac{2}{3}$

   b. $\dfrac{3}{4}$

   c. $\dfrac{5}{12}$

   d. $\dfrac{13}{15}$

6. a. $\frac{1}{9} + \left(\frac{1}{9}\right)\left(\frac{8}{9}\right) + \left(\frac{1}{9}\right)\left(\frac{8}{9}\right)^2 + \left(\frac{1}{9}\right)\left(\frac{8}{9}\right)^3 + \ldots$

b. 1

Chapter 10 Challenge Set

1. a. 2, 4, 7, 11, 16

b. $t_n = t_{n-1} + n$

2. a. $\frac{1}{2}, \frac{2}{3}, \frac{3}{4}, \frac{4}{5}$

b. $\frac{n}{n+1}$; $\frac{1}{1 \cdot 2} + \frac{1}{2 \cdot 3} + \frac{1}{3 \cdot 4} + \ldots + \frac{1}{n(n+1)} =$

$\left(1 - \frac{1}{2}\right) + \left(\frac{1}{2} - \frac{1}{3}\right) + \left(\frac{1}{3} - \frac{1}{4}\right) + \ldots +$

$\frac{1}{n} - \frac{1}{n+1} = 1 - \frac{1}{n+1} = \frac{n}{n+1}$

c. Cancellations cause the sum to collapse like a nautical telescope.

3. a. The $4 \times 4 \times 4$ cube can be arranged in the following shape, which creates a larger square:

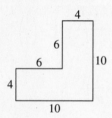

b. $\frac{n(n+1)}{2}$

c. $\left(\frac{n(n+1)}{2}\right)^2$

4. The points corresponding to the first sequence lie on a spiral that expands outward in the counterclockwise direction ($|r| > 1$). The points corresponding to the second sequence traverse the unit circle counterclockwise ($|r| = 1$). The points corresponding to the third sequence lie on a spiral that contracts inward in the counterclockwise direction ($|r| < 1$). These patterns are unaffected by a change in the value of a (the spirals start at different points; the circle may not have radius 1). Using conjugates changes the direction of motion to clockwise.

Chapter 11

Challenge Set 60

1. a. $y = -a(x - a) + 1$

b. Distance from $(x, -a(x - a) + 1)$ to $(0, 0) =$
$\sqrt{x^2 + a^2(x - a)^2 - 2a(x - a) + 1}$; distance from $(x, -a(x - a) + 1)$ to $(2a, 2) =$
$\sqrt{(x - 2a)^2 + (-a(x - a) - 1)^2} =$
$\sqrt{x^2 - 4ax + 4a^2 + a^2(x - a)^2 + 2a(x - a) + 1}$
$= \sqrt{x^2 + a^2(x - a)^2 - 2a(x - a) + 1}$

2. a. $\frac{c}{b-a} \cdot \frac{c}{b+a} = -1$; $c^2 = a^2 - b^2$

b. $[(b + a)^2) + c^2] + [(b - a)^2 + c^2] = 4a^2$;
$2(b^2 + a^2 + c^2) = 4a^2$; $c^2 = a^2 - b^2$

c. This shows that two lines are perpendicular if and only if the product of their slopes is -1.

3. a. $M(a, 0)$, $N(a + b, c)$, $P(b + d, c + e)$, $Q(d, e)$

b. Slopes of $\overline{MN}$ and $\overline{PQ}$ are both $= \frac{c}{b}$; slopes of $\overline{PN}$ and $\overline{QM}$ are both $\frac{e}{d - a}$; the segments joining the midpoints of the sides of a quadrilateral form a parallelogram.

4. a. Slope of $\overline{AC} = \frac{2c}{2b} = \frac{c}{b}$;

slope of $\overline{BD} = \frac{2e}{2d - 2a} = \frac{e}{d - a}$

b. a rectangle

5. a. $PQ = \sqrt{(x - a)^2 + (x - b)^2} =$
$\sqrt{2x^2 - 2(a + b)x + a^2 + b^2}$; the quadratic function under the radical is minimized on the axis of symmetry of the corresponding parabola $x = \frac{a+b}{2}$; therefore, $y = \frac{a+b}{2}$, also.

b. Slope of $\overline{PQ} = \frac{\frac{(a+b)}{2} - a}{\frac{(a+b)}{2}} - b = \frac{b-a}{a-b} = -1$;

slope of $l = 1$; the two lines are perpendicular.

Challenge Set 61

1. $(1, 2)$, $(-2, 11)$

2. $(4, 5)$, $(6, 1)$

3. $(-1, -3)$, $(3, 29)$

4. $(2, 3)$, $\left(-\dfrac{7}{2}, \dfrac{1}{4}\right)$

5. $(9, 116)$, $(-1, 6)$

6. $\left(\dfrac{3}{2}, -\dfrac{1}{2}\right)$, $\left(\dfrac{9}{2}, \dfrac{47}{2}\right)$

7. a. $\dfrac{x^2 - a^2}{x - a}$

 b. $x + a$; $2a$

 c. As Q gets closer and closer to P, the slope
 of the line PQ more nearly approximates the
 slope of the tangent line l.

8. a. $F = \left(0, \dfrac{1}{4}\right)$; $D = \left(a, -\dfrac{1}{4}\right)$

 b. $y - a^2 = 2a(x - a)$

 c. $y - \dfrac{1}{4} = -\dfrac{1}{2a}x$

 d. The slopes are $-\dfrac{1}{2a}$ and $2a$, whose product is

 -1; midpoint of $\overline{FD}$ is $\left(\dfrac{a}{2}, 0\right)$, which is on l.

 The tangent l bisects $\angle FPD$.

Challenge Set 62

1. $(\pm 3, -4)$, $(\pm 4, 3)$

2. $(0, 17)$, $(\pm 8, -15)$

3. $(5, \pm 5)$

4.

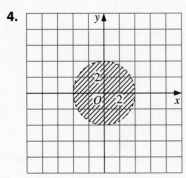

5.

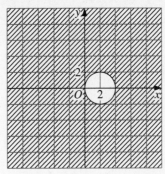

6.

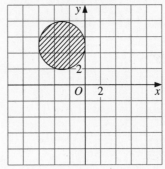

7. $x^2 + y^2 \geq 4$, $x^2 + y^2 \leq 9$

8. a. $\sqrt{x^2 + y^2}$; $\sqrt{(x - 6)^2 + y^2}$

 b. $\sqrt{x^2 + y^2} = 2\sqrt{(x - 6)^2 + y^2}$; $x^2 + y^2 = 4(x^2 - 12x + 36 + y^2)$; $3x^2 + 3y^2 - 48x + 144 = 0$; $x^2 + y^2 - 16x + 48 = 0$; $(x - 8)^2 + y^2 = 16$; center $(8, 0)$; radius $= 4$

9. a. $\left(\dfrac{a + r}{2}, \dfrac{b}{2}\right)$; equation of $\overline{OM}$; $y = \dfrac{b}{a + r}x$

 b. $a^2 + b^2 = r^2$

 c. (slope $\overline{AB}$) • (slope $\overline{OM}$) $= \dfrac{b}{a - r} \cdot \dfrac{b}{a + r} =$

 $\dfrac{b^2}{a^2 - r^2} = \dfrac{b^2}{a^2 - (a^2 + b^2)} = \dfrac{b^2}{-b^2} = -1$; therefore,

 the lines are perpendicular.

10. $(x - 2)^2 + (y + 3)^2 = 25$

11. $(x + 4)^2 + (y - 5)^2 = 169$

Challenge Set 63

1. $(0, -3)$, $\left(-\dfrac{4}{3}, 1\right)$

2. $(2, 0)$, $(1, -3)$

3. $(0, 6)$, $(-5, 4)$

4. $\left(2\sqrt{2}, \frac{3}{2}\sqrt{2}\right), \left(2\sqrt{2}, -\frac{3}{2}\sqrt{2}\right), \left(-2\sqrt{2}, \frac{3}{2}\sqrt{2}\right),$
$\left(-2\sqrt{2}, -\frac{3}{2}\sqrt{2}\right)$

5. a. $\sqrt{(x-3)^2 + y^2} = \frac{3}{5}\left(x - \frac{25}{3}\right)^2$

b. $x^2 - 6x + 9 + y^2 = \frac{9}{25}\left(x^2 - \frac{50}{3}x + \frac{625}{9}\right);$

$\frac{16}{25}x^2 + y^2 = 16; \frac{x^2}{25} + \frac{y^2}{16} = 1; 8$

6. a.

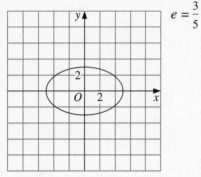

$e = \frac{3}{5}$

b.

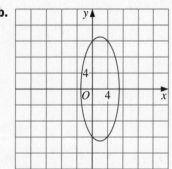

$e = \frac{12}{13}$

c.

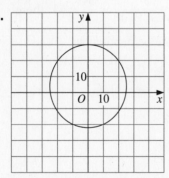

$e = \frac{\sqrt{51}}{26}$

d. The ellipse is more elongated if e is closer to 1 or more circular if e is closer to 0.

7. a. As e approaches 0, a and b become closer in value.

b. As e approaches 0, a and b approach a common value r; as they get closer to r, the expression πab approaches πr^2, the formula for the area of a circle.

Challenge Set 64

1. $\frac{(x-5)^2}{4} - (y+3)^2 = 1; (5 + \sqrt{5}, -3),$
$(5 - \sqrt{5}, -3)$

2. $\frac{(y-5)^2}{4} - \frac{(x-2)^2}{9} = 1; (2, 5 + \sqrt{13}),$
$(2, 5 - \sqrt{13})$

3.

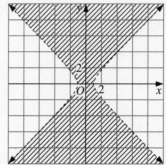

4.

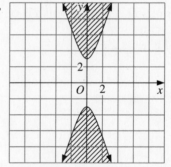

5.

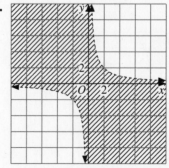

6. a.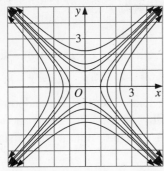

b. They all have the same asymptotes, $y = \pm x$. As $|k|$ approaches $+\infty$, the two branches of the hyperbola become more pointed and approach the asymptotes more gradually.

7. a.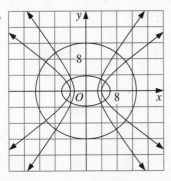

b. They all have the same foci, $(-5, 0)$ and $(5, 0)$; they intersect at right angles.

8. a. $\sqrt{(x+5)^2 + y^2} - \sqrt{(x-5)^2 + y^2} = 6$

b. $\sqrt{(x+5)^2 + y^2} = \sqrt{(x-5)^2 + y^2} + 6$;
$x^2 + 10x + 25 + y^2 = x^2 - 10x + 25 + y^2 + 12\sqrt{(x-5)^2 + y^2} + 36$; $20x - 36 = 12\sqrt{(x-5)^2 + y^2}$; $5x - 9 = 3\sqrt{(x-5)^2 + y^2}$; $25x^2 - 90x + 81 = 9x^2 - 90x + 225 + 9y^2$; $16x^2 - 9y^2 = 144$

c. $\dfrac{x^2}{9} - \dfrac{y^2}{16} = 1$

Challenge Set 65

1. a. 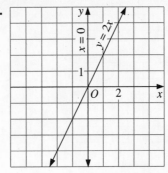 hyperbola

b. 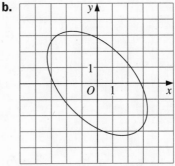 ellipse

c. 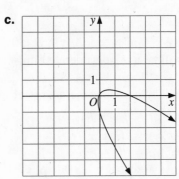 parabola

2. The xy-term causes a rotation of the axes; the axes of the conic section are not parallel or perpendicular to the x- and y-axes.

3. a.

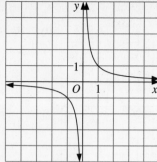

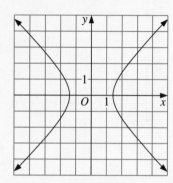

b. vertices: $(1, 1)$, $(-1, -1)$; foci: $(\sqrt{2}, \sqrt{2})$, $(-\sqrt{2}, -\sqrt{2})$

4. a. For $A > 0$, $B > 0$, $C > 0$ or $A < 0$, $B < 0$, $C < 0$, it will be a nondegenerate ellipse. For $A > 0$, $B < 0$, $C \neq 0$ or $A < 0$, $B > 0$, $C \neq 0$, it will be a nondegenerate hyperbola.

b. Let $f(x, y) = A(x - h)^2 + B(y - k)^2$; then $f(h + p, y) = A(h + p - h)^2 + B(y - k)^2 = Ap^2 + B(y - h)^2 = A(h - p - h)^2 + B(y - k)^2 = f(h - p, y)$; therefore $f(h + p, y) = C$ if and only if $f(h - p, y) = C$; the graph is symmetric about the line $x = h$.

c. The proof is almost identical; the graph is symmetric about the line $y = k$.

5. a. hyperbola

b. $(x + y)(x + 2y) = 0$

c.

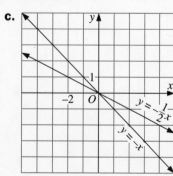

It is the intersection of a double cone with a steeply slanting plane that passes through the vertex of the cone.

Chapter 11 Challenge Set

1. a. $k + c$

b. Solve for x when $y = k + c$: $(k + c) - k = \frac{1}{4c}(x - h)^2$; $4c^2 = (x - h)^2$; $\pm 2c = x - h$; $x = h \pm 2c$. Therefore, the two endpoints of the "width" are $(h - 2c, k + c)$ and $(h + 2c, k + c)$. These points are a distance $4c$ apart.

2. a. $y = \pm\frac{b}{a}\sqrt{a^2 - c^2} = \pm\frac{b^2}{a}$

b. Distance $= \frac{b^2}{a} - \left(-\frac{b^2}{a}\right) = \frac{2b^2}{a}$

c. $\frac{2b^2}{a}$; it is the same expression.

3. $\frac{(x - 5)^2}{36} + \frac{y^2}{11} = 1$

4. a. $OP \cdot OQ = \sqrt{a^2 + b^2} \cdot \sqrt{\frac{a^2}{(a^2 + b^2)^2} + \frac{b^2}{(a^2 + b^2)^2}}$

$= \sqrt{a^2 + b^2} \cdot \sqrt{\frac{a^2 + b^2}{(a^2 + b^2)^2}} =$

$\sqrt{a^2 + b^2} \cdot \sqrt{\frac{1}{a^2 + b^2}} = 1$

b. Q is on the second circle if and only if

$F\left(\frac{a^2}{(a^2 + b^2)^2} + \frac{b^2}{(a^2 + b^2)^2}\right) + \frac{Da}{a^2 + b^2} + \frac{Eb}{a^2 + b^2} + A = 0$. This is equivalent to

$A(a^2 + b^2) + Da + Eb + F = 0$.

5. $PF_1 + PF_2 = 4$; $\sqrt{(x + 2)^2 + (y + 2)^2} - \sqrt{(x - 2)^2 + (y - 2)^2} = 4$; $\sqrt{(x + 2)^2 + (y + 2)^2} = \sqrt{(x - 2)^2 + (y - 2)^2} + 4$; squaring both sides gives $x^2 + 4x + 4 + y^2 + 4y + 4 = x^2 - 4x + 4 + y^2 - 4y + 4 + 8\sqrt{(x - 2)^2 + (y - 2)^2} + 16$; $8x + 8y - 16 = 8\sqrt{(x - 2)^2 + (y - 2)^2}$; $x + y - 2 = \sqrt{(x - 2)^2 + (y - 2)^2}$; $x^2 + y^2 + 4 - 4x - 4y + 2xy = x^2 - 4x + 4 + y^2 - 4y + 4$; $2xy = 4$; $xy = 2$

Chapter 12

Challenge Set 66

1. a.

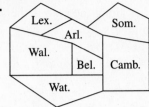

b. Answers may vary. A sample is given. Watertown, Somerville, and Lexington: red; Cambridge and Waltham: green; Belmont: blue; Arlington: yellow

c. The map can be colored with 3 colors; for example, Waltham, Cambridge: red; Watertown, Arlington: green; Lexington, Belmont, Somerville: yellow

2. a.

$$
\begin{array}{c}
\quad A\ B\ C\ D \\
\begin{array}{c} A \\ B \\ C \\ D \end{array}
\left[\begin{array}{cccc}
0 & 1 & 1 & 1 \\
1 & 0 & 1 & 1 \\
1 & 1 & 0 & 0 \\
1 & 1 & 0 & 0
\end{array}\right]
\end{array}
\qquad
\begin{array}{c}
\quad A\ B\ C\ D\ E \\
\begin{array}{c} A \\ B \\ C \\ D \\ E \end{array}
\left[\begin{array}{ccccc}
0 & 0 & 1 & 0 & 1 \\
0 & 0 & 0 & 1 & 1 \\
1 & 0 & 0 & 1 & 1 \\
0 & 1 & 1 & 0 & 1 \\
1 & 1 & 1 & 1 & 0
\end{array}\right]
\end{array}
$$

$$
\begin{array}{c}
\quad P\ Q\ R\ S\ T\ V \\
\begin{array}{c} P \\ Q \\ R \\ S \\ T \\ V \end{array}
\left[\begin{array}{cccccc}
0 & 0 & 0 & 1 & 1 & 0 \\
0 & 0 & 1 & 0 & 1 & 1 \\
0 & 1 & 0 & 1 & 1 & 0 \\
1 & 0 & 1 & 0 & 1 & 0 \\
1 & 1 & 1 & 1 & 0 & 1 \\
1 & 1 & 0 & 0 & 1 & 0
\end{array}\right]
\end{array}
$$

b.

	Wlm	Wtn	A	L	B	S	C
Wlm	0	1	1	1	1	0	0
Wtn	1	0	0	0	1	0	1
A	1	0	0	1	1	1	1
L	1	0	1	0	0	0	0
B	1	1	1	0	0	0	1
S	0	0	1	0	0	0	1
C	0	1	1	0	1	1	0

c. The matrix is symmetric because if, for example, B is connected to A, then A must be connected to B.

3.

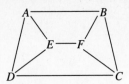

Challenge Set 67

1. a.

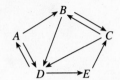

b.
$$
\left[\begin{array}{ccccc}
4 & 4 & 5 & 7 & 4 \\
4 & 5 & 4 & 7 & 4 \\
4 & 4 & 5 & 7 & 4 \\
3 & 4 & 4 & 6 & 3 \\
2 & 3 & 2 & 4 & 2
\end{array}\right]
$$

2. a. 35,000; 15,000; 31,250; 18,750

b. $A^{20} \approx \begin{bmatrix} 0.4019 & 0.3987 \\ 0.5981 & 0.6013 \end{bmatrix}$, $F_{20} \approx 20{,}063$,

$S_{20} \approx 29{,}937$; $A^{40} \approx \begin{bmatrix} 0.4000 & 0.4000 \\ 0.6000 & 0.6000 \end{bmatrix}$,

$F_{40} \approx 20{,}000$, $S_{40} \approx 30{,}000$

c. The ratio between the populations of the towns tends toward 2:3; no.

3. a. $C_{n+1} = 0.4C_n + 0.3R_n$; $R_{n+1} = -0.4C_n + 1.2R_n$; $A = \begin{bmatrix} 0.4 & 0.3 \\ -0.4 & 1.2 \end{bmatrix}$

b. $C_{40} = -0.5C_0 + 0.75R_0$; $R_{40} = -C_0 + 1.5R_0$. The coyote population tends toward $\frac{1}{2}$ the rabbit population.

c. Both populations tend toward 0; mutual extinction.

Challenge Set 68

1. a. 86,400

b. 2,073,600

2. a. 7,200,000

b. 7,160,000

3. a. 19,509,360

b. 17,316,000

4. 390,600

5. a. 524,288

   b. 1,310,720

6. 35; 70

7. 16

Challenge Set 69

1. 9,834,496

2. a. 756,756

   b. 126,126

3. a. 60

   b. 7200

4. 5148

5. 54,912

6. 123,552

7. 378

8. a. 20

   b. 2, 3, 4, 5, 6, 7 or 8 in.; 1, 6, 15, 20, 15, 6, 1

Challenge Set 70

1. Since $k = {}_nC_r = \dfrac{n!}{(n-r)!\,r!} =$

$n(n-1) \cdots \dfrac{(n-r+1)}{r!}$, multiplying k by $n-r$

and dividing by $r+1$ gives $n(n-1) \cdots$

$\dfrac{(n-r+1)(n-r)}{r!(r+1)} = \dfrac{n!}{(n-r-1)!\,(r+1)!} = {}_nC_{r+1}$

2. 21

3. 10

4. 15

5. 56

6. a. $a^5 + 5a^4b + 10a^3b^2 + 10a^2b^3 + 5ab^4 + b^5$

   b. By repeated application of the Distributive Property, you get a sum of terms with five a- and/or b-factors each, with each term being repeated according to the number of ways it can be produced by multiplying the appropriate number of copies of a and b in different orders.

c. You are choosing r factors out of 5 from which to pick the copies of b, for example. There are ${}_5C_r$ ways to do this.

7. a. The last number along the path is the sum of the other numbers.

   b. ${}_5C_3 = \displaystyle\sum_{k=0}^{3} {}_{k+1}C_k;\ {}_5C_2 = \sum_{k=0}^{2} {}_{k+2}C_k;$

$${}_{r+p+1}C_r = \sum_{k=0}^{r} {}_{k+p}C_k$$

8. a. ${}_nC_r + {}_nC_{r+1} = {}_{n+1}C_{r+1}$

   b. ${}_nC_r + {}_nC_{r+1} = \dfrac{n!}{(n-r)!\,r!} + \dfrac{n!}{(n-r-1)!\,(r+1)!} =$

$\dfrac{(r+1)n! + (n-r)n!}{(n-r)!(r+1)!} = \dfrac{(n+1)n!}{(n-r)!(r+1)!} =$

$\dfrac{(n+1)!}{((n+1)-(r+1))!(r+1)!} = {}_{n+1}C_{r+1}$

Chapter 12 Challenge Set

1. 11

2. 10

3. 9

4. a. $8! = 40,320$

   b. 8; $7! = 5040$

   c. $(n-1)!$

5. Since ${}_nC_r = n(n-1) \cdots \left(\dfrac{n-r+1}{r!}\right)$, the

numerator always contains the factor n for $r > 0$, $r < n$. If n is prime, no factor of n can cancel with the denominator.

6. a. $1 + \dfrac{1}{2}x + \left(-\dfrac{1}{8}\right)x^2 + \dfrac{1}{16}x^3$

   b. 1.118; this is accurate to 3 decimal places.

   c. $2 + \dfrac{1}{12}x - \dfrac{1}{288}x^2 + \dfrac{5}{20{,}736}x^3$; 2.0801; this is accurate to 4 decimal places.

Chapter 13

Challenge Set 71

1. $\dfrac{1}{9}; \dfrac{1}{3}; \dfrac{5}{9}$

2. a–b.

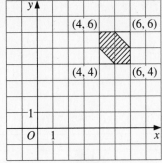

(4, 6)   (6, 6)

(4, 4)   (6, 4)

c. $\dfrac{3}{4}$

3. a. $\dfrac{5}{72}$

b. $\dfrac{5}{12}$

4. a.

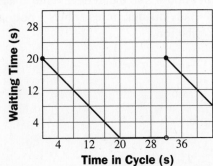

b. Each y-value represents a time you will have to wait. The average should be the "sum" of the y-values (the area under the graph) divided by the "number" of possible arrival times (the length of the interval); $\dfrac{25}{4}$ s

5. a. 2(rand)

b. Square each random number; find the ratio p of the number of random numbers whose square is less than or equal to 2 to the total number of random numbers; p is theoretically $\dfrac{\sqrt{2}}{2}$. Therefore, $2p$ will approximate $\sqrt{2}$.

6. a. 2; π

b. Find the ratio p of darts landing in the unshaded region to darts landing in the semicircle BEC. This ratio should be $\dfrac{2}{\pi}$. Therefore, $\pi \approx \dfrac{2}{p}$.

Challenge Set 72

1. a. 0.4

b. 42%

c. 58%

2. a. about 0.63

b. 13

3. a. $\dfrac{2}{11}$

b. $\dfrac{3}{11}$

c. $\dfrac{5}{11}$

d. $\dfrac{6}{11}$

4. a. The probability that they will be of opposite colors, $\dfrac{26}{51}$, is greater; the probability that they will be the same color is $\dfrac{25}{51}$.

b. If the population is $2n$, the probability that they are of the same party is $\dfrac{n-1}{2n-1}$, which is very close to $\dfrac{1}{2}$ if n is large, so the probabilities are about equal. However, the probability that they will be of opposite parties is slightly greater.

5. 48.5%

Challenge Set 73

1. a. A and B are independent if and only if $P(A \text{ and } B) = P(A) \cdot P(B)$; but $P(A \text{ and } B) = P(A) \cdot P(B \mid A)$; therefore, $P(A) \cdot P(B) = P(A) \cdot P(B \mid A)$, so $P(B) = P(B \mid A)$.

b. $P(B) = P(B \mid A) \Leftrightarrow \dfrac{b}{d} = \dfrac{c}{a} \Leftrightarrow ab = cd \Leftrightarrow$ $\dfrac{a}{d} = \dfrac{c}{b} \Leftrightarrow P(A) = P(A \mid B)$.

2. a. $\dfrac{5}{7}$

b. $\dfrac{7}{10}$

3. a. $\dfrac{2}{5}$

b. $\dfrac{2}{3}$

c. $\frac{2}{3}$

d. $\frac{8}{15}$

e. $\frac{3}{5}$

4. a. $\frac{2}{27}$

b. $\frac{5}{9}$

c. $\frac{1}{9}$

d. $\frac{56}{81}$

5. a. $\frac{1}{2}$

b. $\frac{1}{3}$

Challenge Set 74

1. a. $\frac{3}{8}$

b. $\frac{2}{3}$

2. a. 0.36; 0.16; 0.48

b. $(0.48)^n$

c. 14 points

3. a. $36p^{10} - 80p^9 + 45p^8$

b. $p > 0.74$

4. a. $\sum_{k=0}^{3} {}_3C_k\, p^k(1-p)^{3-k} = p^3 + 3p^2(1-p) +$
$3p(1-p)^2 + (1-p)^3 = p^3 + 3p^2 - 3p^3 +$
$3p - 6p^2 + 3p^3 + 1 - 3p + 3p^2 - p^3 = 1$

b. The probability of 0, 1, 2, or 3 successes in 3 binomial trials is 1.

c. $\sum_{k=0}^{n} {}_nC_k\, p^k(1-p)^{n-k} = 1$

5. $\frac{5}{16}$; $\frac{37}{256}$; the probability gets smaller as n gets larger; it is approximately halved every time n is increased by 1.

Challenge Set 75

1. 0.5 oz

2. a. 7.5

b. 10

c. 0.130

3. a. 1, 10, 45, 120, 210, 252, 210, 120, 45, 10, 1;

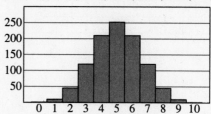

b. 1.58

c. About 66%; the theoretical value is about 68%; the histogram resembles a normal distribution.

4. a. -3; σ

b. 0; $\frac{\sigma}{5}$

c. 3.5; $\frac{\sigma}{2}$

5. $-1.7, 0.7$

6. This factor causes the area under the graph from $-\infty$ to $+\infty$ to be 1; without it, the area would $\sqrt{2\pi}$.

Chapter 13 Challenge Set

1. a. $\frac{364}{365}$

b. $\frac{363}{365}$, or about 0.99

c. 15

2. a. $\frac{8}{15}$

b. $\frac{1}{2}$

c. $\frac{4}{15}$; no

3. 0.54

4. a. $\text{int}(2 * \text{rand}) - 1$

b.

Sum	Probability
4	$\frac{1}{16}$
2	$\frac{3}{16}$
0	$\frac{6}{16}$
–2	$\frac{3}{16}$
–4	$\frac{1}{16}$

5. $\frac{7}{8}$; for $-10 \le x \le 0$, (area between -10 and x)
$= \frac{(x + 10)^2}{200}$; for $0 \le x \le 10$, (area between -10
and x) $= 1 - \frac{(10 - x)^2}{200}$

Chapter 14

Challenge Set 76

1. a. The quantity equals 1 for all A. $\sin^2 A + \cos^2 A = 1$ for all acute angles.

b. $\sin A = \frac{a}{c}$; $\cos A = \frac{b}{c}$; $\sin^2 A + \cos^2 A =$
$\left(\frac{a}{c}\right)^2 + \left(\frac{b}{c}\right)^2 = \frac{(a^2 + b^2)}{c^2} = \frac{c^2}{c^2} = 1$

2. 0.6018, 0.6018; 0.9336, 0.9336;
$\sin \theta = \cos (90 - \theta)$; $\sin A = \frac{a}{c} = \cos B$,
$m\angle B = 90° - m\angle A$

3. a. $\triangle PQS \sim \triangle PRQ$

b. $\frac{x}{1} = \frac{1}{1 + x}$; $x^2 + x - 1 = 0$; $x = \frac{-1 \pm \sqrt{5}}{2}$; only
the positive root makes sense as a length.

c. $\sin 18° = \frac{-1 + \sqrt{5}}{4} \approx 0.3090$; yes

4. a. $40°$; $x = 2 \sin \frac{\theta}{2}$; perimeter $= 18 \sin \frac{\theta}{2} =$
$18 \sin 20°$

b. $\theta = \frac{360}{n}$; perimeter $= 2n \sin \frac{180}{n}$

c. 6.26; 36 sides (perimeter ≈ 6.28)

Challenge Set 77

1. a. The value of the expression is always 1;
$\sec^2 A - \tan^2 A = 1$

b. $\sec^2 A - \tan^2 A = \frac{c^2}{b^2} - \frac{a^2}{b^2} = \frac{(c^2 - a^2)}{b^2} = \frac{b^2}{b^2} = 1$

2. 130 ft

3. $h = \dfrac{c}{\dfrac{1}{\tan A} + \dfrac{1}{\tan B}}$ or $h = \dfrac{c \tan A \tan B}{\tan A + \tan B}$

4. 0.74 mi

5. a. A tangent is perpendicular to the radius drawn to the point of tangency.

b. *PA*; *OA*

c. The segment whose length is $\tan A$ is part of a tangent line to the circle; the segment whose length is $\sec A$ is part of a secant line.

6. $\dfrac{x}{\sqrt{1 - x^2}}$

Challenge Set 78

1. 3.571; 1.042; 0.2917

2. 1.667; –1.25; –0.75

3. –1.970; –1.161; 0.5893

4. –1.45; 1.381; –0.9524

5. $\cos \theta = -\frac{40}{41}$; $\tan \theta = -\frac{9}{40}$

6. $\sin \theta = -\frac{39}{89}$; $\cos \theta = \frac{80}{89}$

7. $\sin \theta = -\frac{48}{73}$; $\tan \theta = \frac{48}{55}$

8. a. $\tan \theta = \frac{\sin \theta}{\cos \theta}$; $\cot \theta = \frac{\cos \theta}{\sin \theta}$

b. $\tan \theta + \cot \theta = \frac{\sin \theta}{\cos \theta} + \frac{\cos \theta}{\sin \theta} = \frac{\sin^2 \theta + \cos^2 \theta}{\sin \theta \cos \theta} =$
$\frac{1}{\sin \theta \cos \theta} = \sec \theta \csc \theta$

9. a. They are equal.

b. $\tan^2 \theta - \sin^2 \theta = \tan^2 \theta \sin^2 \theta$;

Proof: $\tan^2 \theta - \sin^2 \theta = \dfrac{\sin^2 \theta}{\cos^2 \theta} - \sin^2 \theta =$

$\dfrac{\sin^2 \theta - \sin^2 \theta \cos^2 \theta}{\cos^2 \theta} = \dfrac{\sin^2 \theta(1 - \cos^2 \theta)}{\cos^2 \theta} =$

$\sin^2 \theta \cdot \dfrac{\sin^2 \theta}{\cos^2 \theta} = \sin^2 \theta \tan^2 \theta$

10. a. $P(r \cos \theta, r \sin \theta)$;
   $Q(r \cos \theta + s \cos \phi, r \sin \theta + s \sin \phi)$

   b. 21 mi

Challenge Set 79

1. a. The right side is the fraction of the circumference represented by the angle of the sector.

   b. Area $= \dfrac{1}{2}rL$

   c. 3 in., 4 in.

2. a.

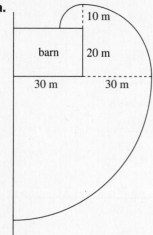

   b. 1150π

3. a.

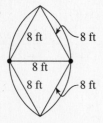

   b. $\dfrac{128\pi}{3} - 32\sqrt{3}$

4. a. $\tan \theta = \dfrac{h}{x}$; $\tan \phi = \dfrac{h}{c - x}$

b. $x = \dfrac{h}{\tan \theta}$; $\tan \phi = \dfrac{h}{\frac{c-h}{\tan \theta}} = \dfrac{h \tan \theta}{c \tan \theta - h}$;

   $(\tan \phi)(c \tan \theta - h) = h(\tan \theta)$;
   $c \tan \phi \tan \theta = h(\tan \phi + \tan \theta)$;

   $h = \dfrac{c \tan \phi \tan \theta}{\tan \phi + \tan \theta}$

c. Area $= \left(\dfrac{1}{2}\right)\dfrac{c^2 \tan \phi \tan \theta}{\tan \phi + \tan \theta}$

Challenge Set 80

1. a. From the law of sines: $\dfrac{b}{\sin B} = \dfrac{a}{\sin A}$;

   $b = \dfrac{a \sin B}{\sin A}$

   b. Area $= \dfrac{1}{2}ab \sin C = \left(\dfrac{1}{2}\right)a^2(\sin B)\dfrac{\sin C}{\sin A}$

2. $\sin P = \dfrac{3}{5}$; $\sin Q = \dfrac{15}{17}$; $q = \dfrac{150}{17} \approx 8.82$

3. Since $\tan A < 0$, A must be obtuse. The law of sines cannot give two solutions in this case.

4. Another solution for $m\angle B$ is $180 - 65 = 115°$, since $\sin 65° = \sin 115°$. The angles are $23°$, $65°$, $92°$ and $23°$, $115°$, $42°$

5. 212 m

6. a. $\sin C = \sin (180 - (A + B)) = \sin (A + B)$, since for any angle θ, $\sin (180 - \theta) = \sin \theta$;

   therefore, by the law of sines, $\dfrac{\sin C}{c} =$

   $\sin \dfrac{A + B}{c} = \dfrac{\sin B}{b}$; so, $\sin (A + B) = \dfrac{c \sin B}{b}$

   b. $c = x + y = b \cos A + a \cos B$; therefore,

   $\sin (A + B) = \sin B \cos A + \dfrac{a \sin B \cos B}{b}$

   c. From the law of sines, $\dfrac{\sin A}{a} = \dfrac{\sin B}{b}$, which is

   equivalent to $\dfrac{a}{b} = \dfrac{\sin A}{\sin B}$. Substituting the right-

   hand side for $\dfrac{a}{b}$ in the equation from part (b),

   you get $\sin (A + B) = \sin B \cos A + \dfrac{\sin A}{\sin B} \cdot$

   $(\sin B \cos B) = \sin B \cos A + \sin A \cos B$

Challenge Problems, ALGEBRA 2: EXPLORATIONS AND APPLICATIONS

Challenge Set 81

1. a. 6.26, 6.27, 6.28; 2π

   b. The perimeter of the polygon more and more closely approximates the circumference of the circle, which is 2π, as n approaches ∞. Draw isosceles triangles with vertices at the origin, and sides of the polygon as bases. The vertex angle of each such triangle will have measure $\dfrac{360}{n}$. By the law of cosines, applied to these triangles, the given expression represents the perimeter of the polygon.

2. a. $\dfrac{285}{323}$

   b. 50

3. $\sqrt{35} \approx 5.92$

4. a. $\cos(180 - A) = -\cos A$; yes

   b. $x^2 = 164 - 160 \cos A$; $x^2 = 244 + 240 \cos A$

   c. $\cos A = -0.2$; $x = 14$

5. a. $(EG)^2 = a^2 + b^2 - 2ab \cos D$; $(DF)^2 = a^2 + b^2 - 2ab \cos E = a^2 + b^2 + 2ab \cos D$

   b. Adding the respective sides of the two equations in part (a) produces the desired result, since the cosine terms cancel.

Chapter 14 Challenge Set

1. a. In right triangle OPS, $\dfrac{\text{adjacent}}{\text{hypotenuse}} = \dfrac{OS}{1} =$ $\cos \angle POS$. That is, $\cos(\alpha + \beta) = OS$.

   b. $OT = \cos \alpha \cos \beta$

   c. Since $\triangle OSV \sim \triangle PRV$, by AA similarity, $m\angle VPR = \alpha$. That is, $m\angle QPR = \alpha$. $OT = \sin \alpha \sin \beta$.

   d. $\cos(\alpha + \beta) = \cos \alpha \cos \beta - \sin \alpha \sin \beta$

2. a. 77.3

   b. $27.2°$

3. Find x by the law of cosines applied to $\angle Q$ in $\triangle ABQ$. Use the facts that the sum of the angles of a triangle is $180°$ and that $\triangle ABQ$ is isosceles to find $m\angle QAB = m\angle QBA$. Then $m\angle PAB = 180 - m\angle QAB - \alpha$, and $m\angle PBA = 180 - m\angle QBA - \beta$. Then use the law of sines in $\triangle PAB$ to find AP; AP is about 33,000 mi.

Chapter 15

Challenge Set 82

1. a. $\sin(90 + \theta) = \sin 90 \cos \theta + \sin \theta \cos 90 = 1 \cdot \cos \theta + \sin \theta \cdot 0 = \cos \theta$

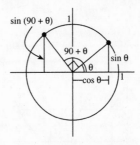

   b. $\sin(180 - \theta) = \sin 180 \cos(-\theta) + \sin(-\theta) \cdot \cos 180 = 0 + (-\sin \theta)(-1) = \sin \theta$

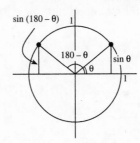

   c. $\sin(\theta - 90) = \sin \theta \cos(-90) + \sin(-90) \cdot \cos \theta = \sin \theta \cdot 0 + (-1) \cos \theta = -\cos \theta$

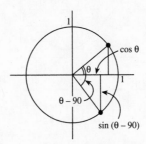

2. a. $\sin(\theta - 45) = \dfrac{\sqrt{2}}{2}(\sin \theta - \cos \theta)$

b.

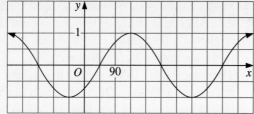

The graph is congruent to the sine graph but translated 45° to the right; it is congruent to the cosine graph but translated 135° to the right.

3. 45°, 135°, 225°, 315°

4. 39°, 141°, 219°, 321°

5. 60°, 120°, 240°, 300°

6. 19°, 161°, 199°, 341°

7. a.

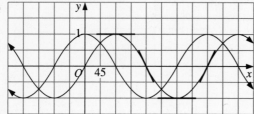

b. cos θ; the slope graph has a zero where the sine graph has a local maximum or a local minimum.

8. a. $8^2 = y^2 + 3^2 - 2 \cdot 3y \cos(90 - \theta)$;
$y^2 - 6y \sin\theta - 55 = 0$

b. $y = \dfrac{6\sin\theta + \sqrt{36\sin^2\theta + 220}}{2} =$
$3\sin\theta + \sqrt{9\sin^2\theta + 55}$

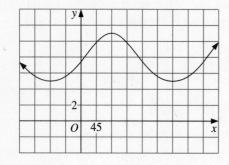

Challenge Set 83

1.

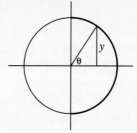

Each value, such as y in the diagram, determines an angle θ between $-\dfrac{\pi}{2}$ and $\dfrac{\pi}{2}$. For cosine, you could choose the range $0 \le \theta \le \pi$; then each x determines a unique θ.

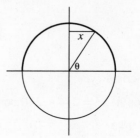

2. a. $-\dfrac{\pi}{6}$

b. $\dfrac{\pi}{4}$

c. $\dfrac{3\pi}{4}$

d. $-\dfrac{\pi}{2}$

3. $\dfrac{\pi}{3}, \dfrac{5\pi}{3}, \pi$

4. $\dfrac{\pi}{3}, \dfrac{2\pi}{3}, \dfrac{4\pi}{3}, \dfrac{5\pi}{3}$

5. $0, \pi, \dfrac{7\pi}{6}, \dfrac{11\pi}{6}$

6. $\dfrac{\pi}{4}, \dfrac{3\pi}{4}, \dfrac{5\pi}{4}, \dfrac{7\pi}{4}$

7. a. 0.985, 0.993, 0.998, 1.000

b. The values of the function $\dfrac{\sin\theta}{\theta}$ approach 1 as θ approaches 0.

8. a. The corresponding values of the two functions are very close.

b.

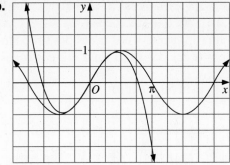

Near $\frac{3\pi}{2}$, for example, the values of the two functions are far apart.

c.

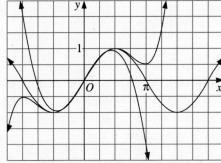

The graph of g stays closer to the graph of $\sin x$ over a wider interval of x-values.

9.

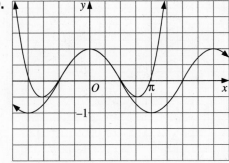

The values of these two functions are fairly close for $-\frac{\pi}{2} \le x \le \frac{\pi}{2}$.

Challenge Set 84

1. a. $\sin 2\theta = \sin (\theta + \theta) = \sin \theta \cos \theta + \sin \theta \cos \theta = 2 \sin \theta \cos \theta$; $\cos 2\theta = \cos (\theta + \theta) = \cos \theta \cos \theta - \sin \theta \sin \theta = \cos^2 \theta - \sin^2 \theta$

b. $\cos 2\theta = 1 - 2 \sin^2 \theta = 2 \cos^2 \theta - 1$

2. $\sin \frac{\gamma}{2} = \pm \sqrt{\frac{1 - \cos \gamma}{2}}$; $\cos \frac{\gamma}{2} = \pm \sqrt{\frac{1 + \cos \gamma}{2}}$;

you can determine the sign by finding the quadrant in which $\frac{\gamma}{2}$ falls.

3. $(\sin \theta + \cos \theta)^2 = \sin^2 \theta + 2 \sin \theta \cos \theta + \cos^2 \theta = 1 + \sin 2\theta$

4. $\dfrac{1}{\cos^2 (\theta/2) + \dfrac{1}{\sin^2 (\theta/2)}} = \dfrac{2}{1 + \cos \theta} + \dfrac{2}{1 - \cos \theta} =$

$\dfrac{2 - 2 \cos \theta + 2 + 2 \cos \theta}{1 - \cos^2 \theta} = \dfrac{4}{\sin^2 \theta}$

5. a.

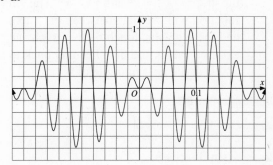

2.5, 25

b. $y = \sin (440\pi t) \sin (1{,}680{,}000\pi t)$

6. a.

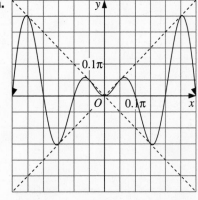

The amplitude of the graph is x, so the waves have peaks and troughs that are progressively farther from the x-axis as you move away from the origin.

b.

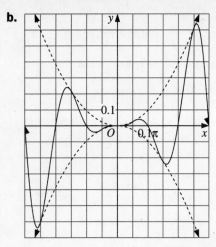

Answers may vary. A sample is given. Draw the graphs of $y = x^2$ and $y = -x^2$. The sine function will oscillate between the x-axis and these graphs, touching the graphs at extreme values and then heading back to the x-axis at the next multiple of 0.1π.

Challenge Set 85

1. $y = \sin \theta + \sqrt{3} \cos \theta$

2. $\dfrac{3\sqrt{2}}{2}(\sin \theta - \cos \theta)$

3. $-\dfrac{5\sqrt{3}}{2} \sin \theta + \dfrac{5}{2} \cos \theta$

4. a.

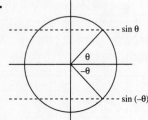

b.

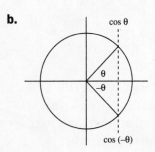

5. a. $\sin (\alpha + \beta) + \sin (\alpha - \beta) = \sin \alpha \cos \beta +$
$\sin \beta \cos \alpha + \sin \alpha \cos (-\beta) +$
$\sin (-\beta) \cos \alpha = \sin \alpha \cos \beta + \sin \beta \cos \alpha +$
$\sin \alpha \cos \beta - \sin \beta \cos \alpha = 2 \sin \alpha \cos \beta$

b. $\cos (\alpha + \beta) + \cos (\alpha - \beta) = \cos \alpha \cos \beta -$
$\sin \beta \sin \alpha + \cos \alpha \cos (-\beta) -$
$\sin (-\beta) \sin \alpha = \cos \alpha \cos \beta - \sin \beta \sin \alpha +$
$\cos \alpha \cos \beta + \sin \beta \sin \alpha = 2 \cos \alpha \cos \beta$

6. $y = \left(\dfrac{1}{2}\right)(\sin 5\theta + \sin \theta)$

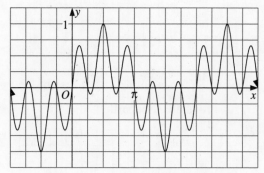

Yes

7. a. Any such point (p, q) is on the unit circle. Therefore, for any angle θ determined by the radius to (p, q) and the positive x-axis, $\cos \theta = p$ and $\sin \theta = q$.

b. $p^2 + q^2 = \dfrac{a^2}{a^2 + b^2} + \dfrac{b^2}{a^2 + b^2} = \dfrac{a^2 + b^2}{a^2 + b^2} = 1$

c. $y = \sqrt{a^2 + b^2} \sin (\alpha + \theta)$

8.

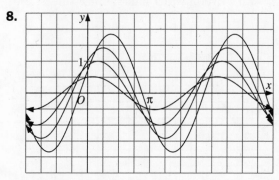

The amplitude shrinks as k gets closer to π.

Challenge Set 86

1. $\tan^2 \theta - \sin^2 \theta = \dfrac{\sin^2 \theta}{\cos^2 \theta} - \sin^2 \theta =$

$\dfrac{\sin^2 \theta - \sin^2 \theta \cos^2 \theta}{\cos^2 \theta} = \dfrac{\sin^2 \theta(1 - \cos^2 \theta)}{\cos^2 \theta} =$

$\sin^2 \theta \cdot \dfrac{\sin^2 \theta}{\cos^2 \theta} = \sin^2 \theta \tan^2 \theta$

2. $\dfrac{\tan^2 \theta}{1 + \tan^2 \theta} = \dfrac{\dfrac{\sin^2 \theta}{\cos^2 \theta}}{1 + \dfrac{\sin^2 \theta}{\cos^2 \theta}} = \dfrac{\dfrac{\sin^2 \theta}{\cos^2 \theta}}{\dfrac{\sin^2 \theta + \cos^2 \theta}{\cos^2 \theta}} = \dfrac{\dfrac{\sin^2 \theta}{\cos^2 \theta}}{\dfrac{1}{\cos^2 \theta}} =$

$\sin^2 \theta$

3. $\tan \theta + \dfrac{1}{\tan \theta} = \dfrac{\sin \theta}{\cos \theta} + \dfrac{1}{\frac{\sin \theta}{\cos \theta}} = \dfrac{\sin \theta}{\cos \theta} + \dfrac{\cos \theta}{\sin \theta} =$

$\dfrac{\sin^2 \theta + \cos^2 \theta}{\sin \theta \cos \theta} = \dfrac{1}{\sin \theta \cos \theta}$

4. $0, \pi, \dfrac{\pi}{4}, \dfrac{5\pi}{4}$

5. $\dfrac{\pi}{6}, \dfrac{5\pi}{6}, \dfrac{7\pi}{6}, \dfrac{11\pi}{6}$

6. a. $\dfrac{1}{\tan 2\theta} = \dfrac{1}{\frac{\sin 2\theta}{\cos 2\theta}} = \dfrac{\cos 2\theta}{\sin 2\theta} = \dfrac{\cos^2 \theta - \sin^2 \theta}{2 \sin \theta \cos \theta} =$

$\dfrac{\cos \theta}{2 \sin \theta} - \dfrac{\sin \theta}{2 \cos \theta} = \left(\dfrac{1}{2}\right)\left(\dfrac{1}{\tan \theta} - \tan \theta\right)$

b. $\tan 2\theta = \dfrac{1}{\frac{1}{\tan 2\theta}} = \dfrac{1}{\left(\frac{1}{2}\right)\left(\frac{1}{\tan \theta} - \tan \theta\right)} =$

$\dfrac{1}{\left(\frac{1}{2}\right)\left(\frac{1 - \tan^2 \theta}{\tan \theta}\right)} = \dfrac{1}{\frac{1 - \tan^2 \theta}{2 \tan \theta}} = \dfrac{2 \tan \theta}{1 - \tan^2 \theta}$

7. a. The graph of $y = \tan \theta$ is increasing in this interval, hence is one-to-one.

b.

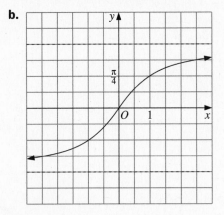

c. y approaches $\dfrac{\pi}{2}$ as x approaches $+\infty$;

y approaches $-\dfrac{\pi}{2}$ as x approaches $-\infty$.

d. Substituting $x = 1$ produces the formula.

8. $\tan^{-1} x + \tan^{-1}\left(\dfrac{1}{x}\right) = \dfrac{\pi}{2}$; in a right triangle with

legs 1 and x, the angle opposite the side of length x has radian measure $\tan^{-1} x$ and the

other acute angle has radian measure $\tan^{-1}\left(\dfrac{1}{x}\right)$.

Since the acute angles of a right triangle are complementary, the result follows.

Chapter 15 Challenge Set

1. a. $\dfrac{\pi}{2} - \dfrac{4}{\pi} \cos x - \dfrac{4}{9\pi} \cos 3x - \dfrac{4}{25\pi} \cos 5x$

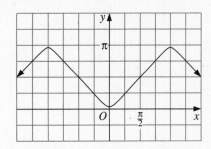

b. $0 = f(0) = \dfrac{\pi}{2} - \dfrac{4}{\pi} \cos 0 - \dfrac{4}{9\pi} \cos 0 -$

$\dfrac{4}{25\pi} \cos 0 - \ldots = \dfrac{\pi}{2} - \dfrac{4}{\pi} - \dfrac{4}{9\pi} - \dfrac{4}{25\pi} - \ldots ;$

$\dfrac{\pi^2}{8} = 1 + \dfrac{1}{9} + \dfrac{1}{25} + \dfrac{1}{49} + \ldots$

2. a. Since $\dfrac{a}{r} = \cos \theta$ and $\dfrac{b}{r} = \sin \theta$, $a = r \cos \theta$; $b = r \sin \theta$; $a + bi = r(\cos \theta + i \sin \theta)$

b–c. $zw = rs[\cos \alpha \cos \beta - \sin \alpha \sin \beta + i(\sin \alpha \cos \beta + \sin \beta \cos \alpha)] = rs[\cos (\alpha + \beta) + i \sin (\alpha + \beta)]$; the magnitude of the product equals the product of the magnitudes of the factors; the angle of the product equals the sum of the angles of the factors.

d. $32[\cos \left(\dfrac{5\pi}{3}\right) + i \sin \left(\dfrac{5\pi}{3}\right)] = 32 - 32i\sqrt{3}$

3.

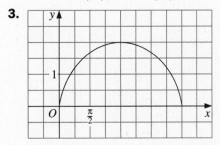

Since $OP = 1$, $\sin t = PQ$ and $\cos t = OQ$. Therefore, $x = t - PQ = t - \sin t$; $y = 1 - OQ = 1 - \cos t$. For any radius r, the equations are $x = r(t - \sin t)$, $y = r(1 - \cos t)$.